Technology Choice and Change in Developing Countries:

Internal and External Constraints

BARBARA G. LUCAS

and

STEPHEN FREEDMAN

Editors

Published by   TYCOOLY INTERNATIONAL PUBLISHING LTD.
DUBLIN

While portions of the research on which Chapters 1, 2 and 3 are based were funded by the National Science Foundation, the opinions, findings and conclusions in this volume are those of the authors and do not necessarily reflect the views of the National Science Foundation.

Published by:
Tycooly International Publishing Ltd.,
6 Crofton Terrace,
Dun Laoghaire,
Co. Dublin, Ireland
Telephone: (+353-1) 800245, 800246
Telex: 30547 SHCN EI

First Edition 1983
© Copyright 1983 Tycooly International Publishing Ltd.

ISBN 0 907567 32 0 Library edition
ISBN 0 907567 33 9 Paperback

Phototypeset by AfricaScience International Publishing Limited, P O B 40047, Nairobi, Kenya.
Printed in Ireland by Irish Elsevier Printers, Shannon.

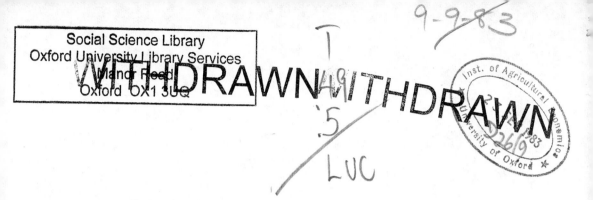
Technology Choice and Change
in Developing Countries:
Internal and External Constraints

CONTENTS

List of Tables

List of Figures

List of Contributors

Editors

BARBARA G. LUCAS, National Science Foundation,
Washington DC

STEPHEN FREEDMAN, Department of Natural Science,
Loyola University of Chicago

Contributors

GUSTAV RANIS, Yale University

GARY SAXONHOUSE, University of Michigan

CARL H. GOTSCH, Stanford University

NORMAN B. McEACHRON, SRI International

JAN SVEJNAR, Assistant Professor of Economics,
Cornell University

ERIK THORBECKE, Professor of Economics and Food Economics,
Cornell University

MARGARET R. BISWAS, Balliol College, Oxford University

ASIT K. BISWAS, President, International Society for Ecological Modelling,
Oxford, England

PIERRE CROSSON, Resources for the Future, Washington

Editor's Introduction

THE IMPORTANCE OF TECHNOLOGY CHOICE and technology change to developing countries has become a perennial concern of development assistance organizations, in international forums, and in development literature. Previous work on the subject of technology choice has provided considerable information on the nature and extent of this process in developing countries.[1] The literature has progressed substantially from theoretical and intuitive discussions on the existence of technology choice to detailed studies of the range of techniques available and their results in terms of investment and labour productivity. Unfortunately, the need for more information about factors influencing the choice and development of more or less appropriate technologies in developing countries continues to outstrip our ability to provide it. In an analysis of 56 national papers about constraints faced by developing countries in using science and technology which were prepared by developing countries for the United Nations Conference on Science and Technology for Development, it was found that over 90 per cent of those papers indicated a need for stronger capabilities in the respective countries for evaluating and selecting among alternative technologies.[2] This volume then is an attempt to add some further information to our understanding of technology choice and change in developing countries.

The papers included in this volume were originally presented at a symposium at the 1980 annual meeting of the American Association for the Advancement of Science. Papers at the symposium were roughly divided into two categories: those presenting preliminary findings from active research on technological behaviour by organizations (presented in Part I of this volume); and papers discussing the importance of considering environmental effects in selecting and developing technologies (presented in Part II).

Part I consists of three papers which discuss some findings of recent research on the determinants of choice of technology by organizations in developing countries. These papers are based on research funded by the National Science Foundation to examine the question of technology choice in developing countries. The Foundation's basic objective was to encourage research which stepped outside of the traditional focus on factor prices to develop and test theoretical frameworks for examining technology choice in an organizational context.[3]

In the first of these studies, Ranis and Saxonhouse (Chapter 1) examine differences in the technological behaviour of Indian and Japanese cotton textile firms in the 19th and 20th centuries and attempt to move beyond simple factor price differences in explaining comparative technological performance. They find that factor price differences and tariff policies were of limited relevance in explaining the markedly different technological behaviour of the Indian and Japanese industries. Rather, the most promising

1

explanations of the industries' different responses to technological opportunities were in the institutional and organizational environments faced by firms in the two countries, which in turn affected the extent of workably competitive pressure faced by firm decision-makers.

The paper written by Gotsch and McEachron (Chapter 2) is also concerned with the impact of government policies and the environments of the firm on the willingness and ability of entrepreneurs to use and adapt more or less appropriate technology. The authors begin by developing a systems model of technology choice. This model is used to examine various components of technological choice and change and to design a representative firm model. The systems model and the firm model are applied empirically to the problem of developing and choosing agricultural machinery in the Pakistan Punjab. The authors examine the technology supplied from both the private and public sectors. They also examine the demand for technology and the linkages between suppliers and users. Next they apply the analytical framework to the generation and choice of agricultural machinery in the Pakistan Punjab to illustrate the potential interaction between the public and private sectors in the generation and diffusion of technology and to emphasize the dynamic character of technology choices in general. Finally they discuss the desirability of focusing future research on linkages and inducement mechanisms on the supply side and provide specific examples.

The Svejnar and Thorbecke paper (Chapter 3) focusses on the design and application of a decision-making model in an attempt to explain the technology choice process through the interaction of different agents. They specify a methodology (social accounting matrix) which can be used to explore and compare the effects of certain technology choices on such macroeconomic objectives as aggregate output, income distribution, and the balance of payments. The possibility of using this approach to identify products (and associated technologies) whose development would benefit particular target groups is also examined. The decision-making model moves from a single-actor, single-goal (profit maximization) model of technology choice to the development of a more realistic multi-actor, multi-goal model where the choice of technology is arrived at through a complex process of bargaining.

Part II begins with a brief introduction by Freedman on the important interrelationships between technological development and the longer term environmental consequences of technological change. His introduction sets the stage for the remaining chapters which stress the importance of including environmental considerations in planning technological development.

The Biswas' paper (Chapter 4) provides a general overview of the problems associated with rationalizing economic development and environmental costs. The authors introduce the topic by presenting a brief narrative history of increasing international concern over environmental protection which led among other things to the United Nations Conference on the Human Environment in 1972. Then, in the final section of the chapter, the authors use the development of the Aswan Dam as a case study from which to illustrate tradeoffs between regional economic development and environmental preservation. The recurring theme presented in this chapter is that development is so closely interrelated with the state of the environment that environmental considerations must be explicitly considered in order for countries to achieve long-term economic growth.

In Chapter 5, Crosson extends the treatment of technology choice by focusing on one form of social cost arising from private decisions about the use of technology. In particular he addresses the problem of controlling environmental damage which results from agricultural production in developing countries, and he considers possible strategies to ensure that environmental costs are included in decisions by farmers and other resource managers. The control of environmental costs is addressed from the twin perspectives of the development of control technologies and the adoption of these technologies by potential users. Crosson suggests that a great deal is known about the techniques for controlling environmental consequences of agricultural production; the difficulty lies in influencing their adoption. Producers are not bearing the environmental costs and, thus, have very little incentive to reduce them.

Traditional approaches to the control of environmental costs of agricultural production are classified into two types—regulatory and incentives approaches. After discussing the problems of equity and implementation associated with these two approaches, Crosson suggests an alternative strategy which he terms the "new technologies strategy". This strategy relies on the "development of new technologies which are at least as attractive to the farmer as those he now employs and which generate significantly lower environmental costs". While the author acknowledges that this strategy could not reduce environmental costs in all cases, it does constitute an improvement over regulatory and incentive strategies in terms of both equity and implementation. Perhaps its major problem is that the development of a "new technologies strategy" is reliant on the environmental sensitivity of public agencies which fund research on agricultural technologies.

The Freedman paper (Chapter 6) focusses on four methods of food production used in developing countries over the last several decades and their respective costs in terms of energy use. Specifically Freedman analyzes the energy efficiency of traditional labour-intensive, capital-intensive and experimental practices in rice cultivation and uses these analyses as a foundation for suggesting specific planning strategies for food production in developing countries. Freedman suggests that by comparing the energy efficiency of individual operations in production methods, it is possible to modify these operations in ways which will increase yields without reducing overall efficiency.

Each of these papers seeks to add in some measure to our understanding of the often overlooked determinants and unavoidable consequences of the choices made by less developed nations as they select technologies to further their economic growth.

Barbara G. Lucas

Notes

[1]See for example the studies cited in A.S. Bhalla, ed., *Technology and Employment in Industry: A Case Study Approach*, with a Foreword by Amartya Sen (Geneva: International Labour Office, 1975); Frances Stewart, *Technology and Underdevelopment* (Boulder, Colorado: Westview Press, 1977) and the special issues of *World Development*, Vol. 5, 1977, Nos. 9 & 10 (combined).

[2]Barbara Lucas, "Working Note: Analysis of UNCSTD Developing Country National Papers" (mimeo) National Science Foundation, August 1979.

[3]The papers contained in Part I of this volume draw on data from the research projects supported by the National Science Foundation; however, the views and interpretations expressed in these papers are those of the authors and should not be attributed to the National Science Foundation or to any individual acting on its behalf.

Part I

Determinants of Choice of Technology

CHAPTER 1

International and Domestic Determinants of Technology Choice by the Less Developed Countries (LDC)*

Gustav Ranis
Yale University
and
Gary Saxonhouse
University of Michigan

FEW SUBJECTS have received more attention in the recent literature on development than that of technology choice and technology change. The recent United Nations Conference on Science and Technology for Development (UNCSTD) in August 1979 has highlighted this concern on an international and political plane; but analysts and policy makers both within the developing countries and outside have for some time recognized the importance of improving the quality of LDC technologies, with direct consequences on productivity and growth and indirect effects aiding the achievement of a more equitable or more employment-intensive growth path. While concern with the role of Science and Technology has been mounting, it is fair to say that the progress in providing answers has been lagging considerably behind. This should not, perhaps, be surprising, in view of the deficiencies in this area of knowledge even in the developed country context. For this reason, any effort to even marginally improve our understanding of how more or less "appropriate" technology is chosen and developed is clearly bound to have a high payoff. This is especially true in the context of societies which are critically short of conventional inputs, a scarcity which lies at the heart of their state of underdevelopment.

That this large payoff exists seems to be indicated by the dramatically divergent performance of a small subset of contemporary developing countries, such as Korea and Taiwan. In the absence of a favourable natural resource

*Paper presented to the Annual Meeting of the American Association for the Advancement of Science in January 1980. Professor Ranis' contribution was financed under a National Science Foundation grant.

endowment, these countries have not only managed to grow rapidly but have avoided a deterioration in the equity of their income distribution over the last quarter century. This record stands in sharp contrast to many other developing countries. Fast growing Latin American countries like Mexico and Colombia, for example, in spite of a more favourable natural resource endowment at the end of World War II, have failed to do as well, either in terms of growth—though this has been respectable—or in achieving distributional, employment and poverty alleviation goals. The inability of these countries to ease painful tradeoffs between output, employment and distributional objectives seems to be associated with considerably less technological flexibility or ability to innovate adaptively in labour-using directions, compared with the Asian example. There are marked differences not only in the intersectoral composition of growth at the aggregative level, with more attention being paid to agriculture at an early point in the case of the Asian countries, but also in the greater scope of labour-intensive technologies afforded by an export oriented industrial output mix. As a consequence, we see in the Asian case a much more rapid absorption of the underemployed and unemployed, accompanied by relatively equitable levels of income distribution, contrasting with worse levels and a deteriorating trend in the Latin American case. Industrial capital-labour ratios seem to be substantially lower in the East Asian countries, and the direction of technology change appears much less labour-saving and thus more "appropriate" than in the majority of contemporary developing countries.

The stakes clearly are high, and thus there is a *prima facie* case for trying to understand the role of more or less appropriate technology in these countries as an important ingredient in an effort to examine the causes of differential performance over the last quarter century. It is a pity that too much of the analysis of technology choice and change has been directed towards the impact of technology on selected aggregate variables of interest, like employment and distribution, with too little analysis of the proximate causes of differential technology performance that are likely to be uncovered by some attempt at disaggregation.

All of this is not intended to deny that we know a good deal more in this general area today than we did a quarter of a century ago. For one, we now recognize the existence, in principle, of a fairly wide range of alternative technology choices for all but a small subset of continuous process industries. For another, we now acknowledge the importance of attempting to include in our analysis the quantitatively important product attribute or quality change along with process change, which is easier for economists to handle. These represent important advances in our understanding. There was little sense in discussing the search for technological alternatives and the inducement mechanism for selecting the appropriate one at the individual firm level, as long as most economists, along with virtually all engineers, still believed in the absolute tyranny of fixed factor proportions and fixed attribute bundles.

As a consequence, today's front lines of research have been moved forward to address the question of what causes more or less appropriate technology choices to be made and what induces a more or less appropriate direction of technology change over time. To put the question in another way, since technological choice can be made among a wide variety of techniques, and since this variety can presumably be further increased through technological change, why is the range of techniques actually used in developed and developing countries so narrow, much narrower than would be expected on the basis of conventional theory, given the international differences in factor endowments?

While we recognize cultural and human resource differentials across countries, it is our basic premise that the environment surrounding economic agents is of crucial significance in answering these questions. Though the supply of information on alternative technologies and appropriate goods remains essential, an active demand by entrepreneurs focussed on the profitability dimension of making more or less appropriate choices is an essential ingredient often neglected in discussions of science and technology. The care that is exercised in technology choice and the level of investment and research activity directed towards technological change depend, to a significant extent, on the type of incentives faced by decision-makers within each firm. Our current research emphasis can be simply summarized by stating that we have selected a number of variables that affect these incentives, and are exploring their effects on technological choice and change in the context of development (Biswas, 1980).

The first of these variables is industrial structure. At one extreme of the industrial structure continuum is agriculture. With the exception of some cash crop plantations, agricultural firms are generally small and competitive. Individual units in agriculture have very weak incentives to invest in research and inventive activity. The legal mechanism of the patent is not effective for most biologically based technology because of the ease of replicability and the difficulty of policing patent protection. The public goods aspect of the generation, adaptation and diffusion of agricultural technology is accordingly prominent. Moving to the industrial sector, we find that the small firms that approximate a competitive industry may also have relatively weak incentives to invest in inventive activity. The presence of many rival firms makes it difficult to amass much market power, even with new technology to retain innovation profits; moreover, their relatively small size may make it difficult to obtain the necessary resources for the pursuit of technology change, and render the impact of potential cost savings less than would be enjoyed by a larger firm. Some of these reasons suggest that firms near the opposite end of the continuum, in fairly concentrated industries, have comparatively strong incentives to conduct research and search activities. However, these inducements can be somewhat offset by the lack of competitive pressure on oligopolistic firms to perform at full potential. Even an industry that is "workably competitive" in structure can be made to feel more or less pressure to innovate, depending on the degree of competition from sources outside the industry, particularly imports. This leads us to the second of the variables we wish to concentrate on, protective measures taken by national governments. We would expect, *ceteris paribus*, firms in industries that are more shielded from foreign competition by tariffs, import quotas and the like, to feel less pressure to innovate and, consequently, to engage in less inventive activity than otherwise. Similarly, firms in industries that are heavily subsidized through credit, foreign exchange rationing, price controls, or other measures, are likely to adopt a less appropriate technology, other things being equal, than firms in unprotected industries.

We have designed our research at this preliminary stage to capture the effects of these two variables through a series of comparisons. We are investigating technological change and choice in agriculture, in a competitive industry, cotton textiles, and in one less competitive industry. Differences between the technological performance of these industries should yield further insights into the effects of private market structure, and the extent and nature of government interventions as part of the environment for individual firm decision-making.

Our total research effort is still in a preliminary stage, one in which we are developing and refining methods capable of testing related hypotheses. This paper presents only preliminary findings on one aspect of our on-going work, that based on an examination of the historical laboratory of the 19th and 20th century Japanese and Indian cotton textile industries.

Section II will focus on the differences in the observed performance with respect to technology in the Indian and Japanese cotton textile industries of the late 19th and early 20th centuries. Section III will examine differences in the economic and institutional environment in the two country settings as possible causal explanations for the observed difference in performance.

DIFFERENTIAL PERFORMANCE

A RELATIVELY ALREADY WELL-ESTABLISHED Indian cotton textile industry and a relatively recent industry in Japan seemed to enjoy substantially similar "initial conditions" at the beginning of the 1880s. The Indian industry, in its modern form, is usually dated from the founding of three successful mills in 1854. By 1882, 65 mills were in operation, housing over 14,000 looms and 1.6 million spindles, almost all of which were mules.[1] Short staple Indian cotton, augmented by about three per cent imported cotton, was spun into relatively coarse, low count yarn.[2] Mill operatives, mostly male, worked a single dawn-to-dusk shift in factories modelled after those in Lancashire[3] (Morris). In Japan, relatively few mills had been erected since the first in 1867. Sixty per cent of yarn consumed was domestically produced with traditional hand spinning methods, and most of the remainder was imported from India, mainly in counts of 20s or higher. As in India, machine spinning was done on mules, mostly by male workers in single shifts.

In the following decade, however, major changes in cotton spinning technology in Japan virtually put an end to this similarity. The data in Table 1, unfortunately available only after 1886, show the general effects of those changes. Their labour-using character is demonstrated by a substantial decline in the capital-labour ratio between 1886-90 and 1891-95. This capital shallowing was combined with an equally dramatic decline in the capital-output ratio. Innovational intensity was clearly large enough to provide an increase in average labour productivity in spite of the strong labour-using bias of the technological changes adopted (Fei and Ranis, 1965). These effects on employment and output are highly desirable in a labour-surplus economy like that prevailing in Japan at this period. Such success is furthermore quite rare in contemporary LDC industrialization experience, and thus warrants our particular attention.

Table 1   The Japanese Spinning Industry: An Aggregative View

Year (average annual)	(1) Capital (average working spindles per day)	(2) Labour (operatives, male & female)	(3) Output (yarn in kan per day)	(4) Capital-Labour Ratio (1) ÷ (2)	(5) Capital-Output Ratio (1) ÷ (3)	(6) Labour Productivity 1000 × (3)÷(2)
1886-90	148,516	5,992	7,887	24.8	18.8	1.32
1891-95	406,419	29,178	42,902	13.9	9.5	1.47
1896-1900	1,013,987	57,857	105,176	17.5	9.6	1.82
1901-05	1,296,471	67,840	120,256	19.1	10.8	1.77
1906-10	1,614,581	80,852	149,419	20.0	10.8	1.85
1911-15	2,331,236	109,228	242,847	21.3	9.6	2.22
1916-19	3,354,972	147,251	303,409	22.8	11.1	2.06

Source: Nippon Teikoku Tokei Nenkan (The Japan Annual Statistical Report) No. 10-40, Japan Cabinet Statistical Bureau, Tokyo. The sharp drop in average annual capital-labour ratio for the industry as a whole between 1886-1890 and 1891-1895 looks suspicious but can be confirmed directly from individual mill records. The capital-labour ratio for the Osaka Spinning Mill dropped from 22.0 in August 1889 to 11.6 in June 1893; for the Naniwa Spinning Mill it dropped from 33.9 to 9.8; for the Kanegafuchi Spinning Mill, from 22.0 to 14.2; and finally for the Mie Spinning Mill from 18.2 to 14.6.

The major innovations and less spectacular adaptations that made up the technological changes of the 1880s can be conveniently divided into two distinct sub-phases. In the first sub-phase, between 1882 and 1884, the spinning mule remained at the centre of the technology, and other changes were made around it. In the second sub-phase, 1887-89, the core process itself was changed as mule spinning was replaced by ring spinning, and some further technological alterations were made in adjusting to rings.

The first significant change, in the first sub-phase, was the adoption of two shifts of eleven hours each, in place of one shift. By this change, the capital-labour ratio in the core spinning processes was reduced by almost half. This sudden and significant intensification of capital use was made possible by greater reliability in motive power, due to a simultaneous change from water power to steam, the increased use of electric lighting, and the availability of sufficient supervisory personnel. The labour in the mills continued to be mostly male at this point.

Beginning in the same period, the emphasis in production was shifted from 16-count yarn to a coarser 12-16 count yarn. This change was important in itself, as finer Indian imports and domestic hand-spun yarn were replaced by low-count yarn with qualities more suitable to the Japanese climate. It also linked up well with improvements that were being made simultaneously in cotton weaving technology through the adoption of Batten- and Jacquard-derived improvements on the traditional hand-loom. These improvements were characterized by the addition of a roller so that the difficult task of the shuttle back and forth by hand as the weft is carried through the warp was considerably eased and made more accessible to lower quality female labour. A second, even more important, improvement, pertaining to the Japanese version of the Batten loom, was the substitution of wood for the metal used in France. This improvement cut capital costs by 50 per cent and accommodated a lower count yarn. Due to the resulting increase in vibrations, the cloth produced was less strong but still suitable in the more important warmth dimensions associated with narrow cloth.

The reduction in yarn count was of even greater significance in preparing the way for the industry's central technological improvement of the decade. Ring spinning, the major alternative to mule spinning, was especially well suited to spinning lower counts. In the short space of two years, 1887-89, virtually the whole industry shifted to the use of rings. The importation of additional mules, excepting a limited number assigned to fine-count production, ceased completely within one year. Aided by some "convenient" fires that destroyed a substantial portion of the existing mule stock, a virtually instantaneous switch from mules to rings could be observed.

This rather dramatic shift in technology in the core spinning process permitted other adaptive technology changes in ancillary processes to take place. The ring machinery clearly had the advantage over mules of accommodating substantially more workers per spindle as well as requiring less skill per worker. Moreover, the ring could be run at higher speeds, and consequently again more labour-intensively, for any given yarn count up to at least a yarn count of 40. Increasingly, young females with a higher average turnover rate were now hired at low wages to provide additional unskilled labour. Moreover, the shift to cotton mixing, itself a labour intensive operation, made it possible to marry the use of the ring to a lower average staple length and additional labour inputs. While cotton mixing had been employed elsewhere, the Japanese were able to avoid a substantial increase in average staple length accompanying the switch from mule to ring because of their own mixing technique innovations.

The overall decline in the capital-labour ratio in Japanese cotton spinning during the second sub-phase was thus mainly related to three factors: the basic shift from mule to ring technology, the increased utilization of cotton mixing, and the policy of machine speed-ups for given two-shift machinery use. These changes augmented the earlier labour-using innovation of the double shift, and continued the emphasis on low-count yarn.

The Indian technological performance over the same period and even later was dramatically different. Given the initial endowment—with at least as much labour surplus—similar types of labour-using changes might have been expected to have taken place in the Indian industry. Table 2 presents an aggregative picture of the changes that occurred. Because Indian statistics for the period do not distinguish between spinning workers and weaving workers in the same mill, it was necessary to estimate the capital-labour ratio for selected years using the average spindle per worker ratio in mills where there were no weaving workers.[4] Such an estimate leads to the conclusion that Indian capital-labour ratios were consistently higher than in Japan and that there was probably no drop in the capital-labour ratio of any size comparable to the Japanese experience, at least in the period for which estimates could be obtained. Thus, while the Indian industry did not become more capital intensive over time, as is customary in contemporary LDC experience, neither did it increase its labour intensity. Instead, it experienced capital-widening, with little apparent movement either in the capital-output ratio or in average labour productivity.

At a time when a new technology was being suggested by British capital suppliers to both Japanese and Indian mills examination at a more disaggregate level leaves the same impression of technological inertness in India. The most striking example is the slow adoption of ring spinning. As Table 3 indicates, the number of rings did increase steadily, and their adoption was viewed as the most significant technical development of the 1880s. Nevertheless, twenty years after their first introduction, 1 million spindles, approximately one-third of total spindleage, were still of the mule variety. Even more significant is the fact that, according to the records of Platt Bros. of Oldham, the main British textile machinery supplier, some 2 million new mules were imported into India between 1883, the date of the first Indian experimentation with rings, and 1900. While this figure unfortunately could not be checked against Indian data, it is clear from Table 3 that large numbers of mules entered the country during this period, in sharp contrast to the sudden halt of Japanese mule imports.

The Indian cotton spinning industry also failed to realize the full labour-using effects of other technological changes as did Japan. Male workers continued to predominate throughout the whole period, with women

Table 2   The Indian Spinning Industry: An Aggregative View

	Ann. Aver. of working spindles (1)	Ann. Av. No. of millhands employed daily (2)	Ann. Av. Yarn output (millions of pounds) (3)	Capital-Labour ratio spindles per worker) (4)	Capital-Output ratio (5)	Aver. Labour Productivity[b] (6)
1895-96—1899-1900	4,210,360	152,406	468.9	32.8	8975.5	3.9
1900-01—1904-05	4,932,343	183,074	571.7	34.7	9275.7	4.0
1905-06—1909-10	5,616,835	221,211	651.6	31.1	8619.9	3.4
1910-11—1914-15	6,257,852	250,739	651.6	30.0	9603.3	3.0
1915-16—1919-20	6,546,723	287,543	669.8[a]	35.1	9774.3	3.5

Source: Columns (1) and (2): *Report of the Indian Tariff Board* (1928), pp. 232-33; column (3): *Statistics of British India* (1921), Vol. 1, p. 59; column (4): computed using the total spindles and total workers in mills with no looms, reported in *Financial and Commercial Statistics of British India* (1897), pp. 416-9 for 1895-96, *ibid.* (1904), pp. 349-50 for 1903-04, *Statistics of British India* (1911) part 1, pp. 36-37 for 1908-09, *Statistical Abstract for British India* (1915), vol. 1, pp. 55-7 for 1913-14, and *Statistics of British India* (1921), pp. 56-8 for 1918-19.

[a]Output for 1919-20 was not available, and so is not included in this average.

[b]The labour averages used were computed by dividing the average number of spindles, (1), by the estimated spindle per worker ratio, (4).

Table 3 Indian Spindleage (millions)

	1884	1894	1914	1923	1929	1939
Mules	2.0*	2.2	1.6	1.1	0.9	0.5
Rings	--	1.2	5.2	6.8	7.9	9.6
Total	2.0	3.4	6.8	7.9	8.8	10.1

Source: Compiled by Toru Yanagihara from Takamura, Kiyokawa and Pearse.

*Including throstles, thus an upper bound figure.

accounting for no more than 25 per cent and children supplying a small and decreasing proportion of the total.[5] Some mills lengthened the work day to fifteen hours in the early part of the century,[6] but even as late as 1930 the single shift was still considered normal.[7] Cotton mixing was certainly practised in India, but its purpose was viewed differently.[8] There is evidence that their mills tried to work with as short a staple length as possible and probably increased their labour use through this practice,[9] yet they did not use mixing to reduce their reliance on short-staple cotton which was less suitable for rings (see Table 3A).

Table 3A Use of Imported Raw Cotton in Indian Cotton Spinning

	Average Annual Cotton Imports to British India (cwt) (1)	Average Annual Indian cotton Consumption (cwt) (2)	Imports as Percentage of Consumption (3)
1880-81 1884-85	51,004	1,652,854	3.1
1885-86 1889-90	73,636	2,837,505	2.6
1890-91 1894-95	89,087	4,256,052	2.1
1895-96 1899-1900	89,377	5,124,087	1.7
1900-01 1904-05	116,422	5,936,090	2.0
1905-06 1909-10	130,937	7,027,437	1.9
1910-11 1914-15	243,637	7,208,410	3.4
1915-16 1919-20	53,876	7,334,676	0.7

Source: Column (1): *Financial and Commercial Statistics,* 4th issue (1897), pp. 518, and *ibid,* 11th issue (1904), pp. 408-09; *Statistics of British India* (1913) 6th issue, vol. I, p. 14; and *ibid,* 1922, 11th issue, vol. 1, p. 138. Column (2): Pearse, *op. cit.,* p. 22.

Table 4   Indian-Japanese Foreign Trade Competition (5-year averages)

Year	Japanese yarn exports to India* (millions of lbs) (1)	Japanese cloth exports to India* (millions of yards) (2)	Japanese yarn exports to China and Hong Kong (millions of lbs) (3)	Indian yarn export to China and Hong Kong (millions of lbs) (4)
1895	--	--	15.3	163.3
1900	--	0.0	88.2	208.4
1905	0.0	0.2	97.9	240.6
1910	0.3	0.2	102.4	171.3
1915	1.4	34.0	170.3	152.4
1920	13.7	133.9	111.1	80.6
1925	26.0	169.2	--	--
1930	8.3	380.6	--	--

Source: Columns (1) and (2): computed from *Annual Statement of Seaborne Trade of British India,* various issues from the 36th to the 64th issue; and *Report of the Indian Tariff Board* (1932), pp. 25 and 28. Columns (3) and (4): computed from *Report of the Indian Tariff Board* (1927), p. 96.

*Amounts for years including parts of two calendar years are treated as if they were for the latter year—e.g. 1904-05 is treated as 1905.

Evidence that the Japanese were forging ahead of the Indians technologically is clearly visible in the relative performance of the two industries when competing for markets. Table 4 shows the increasing penetration of Japanese cloth and yarn into India. In the infancy of the Japanese industry, India had exported yarn to Japan. Despite this early entry, Indian exports to Japan fell, from a peak in 1910-11 of 2.8 million yards of cloth and nearly 3 million pounds of yarn, to less than one per cent of these levels in 1930.[10] The same period saw Japanese yarn entering the Chinese market, which had been mainly supplied by Indian products, and winning the dominant position by 1915 (Table 4). These figures understate the Japanese involvement in China. Japanese-owned mills within China, organized and operated like mills in Japan, expanded their spindleage over eight times between 1915 and 1928, when they accounted for over one-third of total spindleage (Moser, p. 66). The number of Japanese-owned looms increased fifteen times over the same period, and accounted for nearly half the total Chinese weaving capacity in 1928 (*ibid,* p. 87). This expansion makes the Japanese trade victory over India even more convincing than the export figures reveal. Whether in foreign or within the Indian domestic market, the rapid adoption of technology change coupled with the almost instantaneous diffusion discussed earlier gave Japan a decided advantage wherever Indian and Japanese cotton textiles competed.

An explanation of such marked differences in technological behaviour in cotton spinning and weaving will be sought partly in terms of differences in the competitiveness of the macroeconomic environment and partly in differences in the related pattern of final demand constraining the exercise of technological choices. In what follows we make an effort to attribute at least qualitative weights to various plausible causal explanations for the observed difference in technology behaviour.

An Effort at Causal Analysis

The difference in the technological development of the two textile industries is indeed quite startling. Both industries operated in labour surplus economies, both obtained their initial technology and early technical advice from Britain, mostly from the same firm; yet one transformed that technology in ways appropriate to its economy much better than did the other. While it is necessarily difficult, if not impossible, to present any single satisfactory causal explanation of this interesting phenomenon, our objective here is to identify as many *prima facie* plausible connections as possible.

One may suspect that the so-called distortion of relative factor prices inducing entrepreneurs to use too much scarce capital and too little surplus labour was at play in India and less so in Japan. In the absence of data on relative factor availabilities or shadow prices in the two countries it is difficult to make a clear-cut judgment on the initial extent of relative distortion. Inspection of the empirical evidence, however, reveals no apparently striking differences in the trend of relative factor prices over time. Table 5 summarizes the relative movements of wages and the price of capital in Japan. The four price ratios presented suggest that capital was becoming more costly relative to labour up to around 1900, then increasingly less costly. This reversal in trend is consistent with the aggregative endowment picture over time (Ohkawa and Rosovsky, 1968; Ranis, 1959).

Indian evidence in Table 6 presents a somewhat similar picture. The ratios of a rough index of the price of capital to all wages and to cotton textile wages increase or remain constant to about 1900, and decrease thereafter. A similar pattern is seen when an index of interest rates is used to present the price of capital. This does not permit us to conclude that capital wasn't underpriced and labour overpriced in both India and Japan. The nominal wage in India was constant or rose only slowly over the latter half of the nineteenth century (Buchanan; Sarkar). A scatter diagram of the interest rates at which the Indian government borrowed internally shows an overall downward trend through the course of the whole century.[11] The opposite trend occurred in Japan with slower population growth and higher marginal savings rates which tends to give weak support to the hypothesis, i.e. India may have experienced somewhat more credit rationing than Japan. While it is thus difficult to judge differences in the initial extent of factor price distortions, the fact that capital was becoming relatively more expensive in both labour surplus economies over time indicates substantially more factor price flexibility than usually

Table 5   Indices of the Price of Capital and Wages in Japan, in Nominal and Real Terms (1928-32 = 100)

	(1) Capital Goods Price Index	(2) General Wholesale Price Index	(3) Real Price of Capital $\frac{100 \times (1)}{2}$	(4) Money Wage Index	(5) Index of Average Money Wages for Cotton Spinners	(6) Cost of Living Index	(7) Real Wage Index $\frac{100 \times (4)}{6}$	(8) Index of Real Wages of Cotton Spinners $\frac{100 \times (5)}{6}$	(9) Ratio of Real Price of Capital to Real Wages	(10) Ratio of Real Price of Capital to Real Cotton Spinning Wages	(11) Ratio of Capital Goods Price to Nominal Wage Index	(12) Ratio of Capital Goods Price to Nominal Cotton Spinning Wages
1887	29.5	32.2	91.6	11	-	-	-	-	-	-	2.68	-
1888-92	33.4	37.1	90.1	12	9.6	-	-	-	-	-	2.78	3.5
1893-97	43.7	41.7	105.0	14	11.2	32.8	42.7	34.7	2.46	3.0	3.13	3.9
1898-02	59.1	53.1	110.1	18	18.0	44.1	40.8	40.8	2.70	2.7	3.28	3.3
1903-07	63.8	63.5	100.4	19	21.5	50.5	37.6	42.6	2.64	2.4	3.35	3.0
1908-12	65.0	68.5	95.0	27	27.0	55.8	48.4	48.4	1.96	2.0	2.41	2.4
1913-17	86.9	81.2	110.7	29	32.0	60.4	48.0	53.0	2.31	2.1	3.10	2.8
1918-22	170.2	150.4	113.0	74	92.4	119.3	62.0	77.5	1.82	1.5	2.30	1.8
1923-27	148.1	139.6	106.0	108	112.9	118.3	91.3	95.4	1.16	1.1	1.37	1.3
1928-32	100.0	100.0	100.0	100	100.0	100.0	100.0	100.0	1.00	1.0	1.00	1.0
1933-37	100.1	107.4	93.2	93	73.0	98.8	94.1	73.9	.99	1.3	1.08	1.4

Source: G. Ranis, "Factor Proportions in Japanese Economic Development," *American Economic Review*, September 1957 as well as Japanese primary sources.

Table 6 Indices of the Price of Capital, Wages and Interest Rates in India, 1890-1912
(1908-12 = 100)

	Capital Goods Price Index[a] (1)	Index of Average Price Index[b] (2)	Index of All Wages (3)	Index of Cotton Textile Wages (4)	Ratio of Price of Capital to Wages (5) (1) ÷ (3)	Ratio of Price of Capital to Cotton Textile Wages (6) (1) ÷ (4)	Ratio of Interest Rate to Wages (7) (2) ÷ (3)	Ratio of Interest Rate to Cotton Textile Wages (8) (2) ÷ (4)
1890-92	75	71	63	75	1.2	1.0	1.1	0.9
1893-97	79	98	67	77	1.2	1.0	1.5	1.3
1898-02	90	108	76	82	1.2	1.1	1.4	1.3
1903-07	95	99	86	90	1.1	1.1	1.2	1.1
1908-12	100	100	100	100	1.0	1.0	1.0	1.0

Source: Column (1): computed from K.L. Datta, *Report on the Enquiry into the Rise of Prices in India,* Vol. 1. p. 29;
Column (2): *ibid.* Vol. IV, p. 448; Columns (3) and (4): *ibid.,* Vol. III, pp. 2-3.

[a]This index is the average of the price indices for metals and building materials.

[b]This index is the average of the mean annual interest rates on demand loans on Government paper in the Presidency Banks of Bengal, Bombay, and Madras.

encountered in contemporary import substituting LDCs. The relevance of differential factor price distortions to the problem at hand, i.e. explaining the greater capital intensity in India cotton spinning around the turn of the century, is probably limited in any case, as will be made clearer below.[12]

Another frequently encountered source of distortions is, of course, tariff policy. The evidence, however, indicates that Japanese and Indian levels of industrial protection via tariffs during the relevant period were comparably low; in India due to colonial regulations, in Japan as a consequence of the imposed unequal treaty system. Table 7 presents tariff rates for both cotton yarn and cloth for selected years, clearly indicating that, by any current LDC standard, tariffs were extremely low in both industries, particularly during the early infant industry period, with Indian protection somewhat higher in later years, as a direct result of the increased threat of Japanese competition. Whatever relative price distortion was introduced by import duties seems to have been relatively minor and does not appear to explain much of the difference in technological performance.

On the other hand, there are other differences in the organizational/institutional environment facing the two industries which may provide more of an explanation than the old chestnut of relative factor price distortions in its straightforward simple-minded version. We are referring here, in the first instance, to differences in the extent of workable competition as reflected, for example, in the relative freedom of entry, access to credit and pressure in the

Table 7   Tariff Rates on Cotton Yarn and Cotton Cloth in Japan and India, Selected Years

| | Yarn | | Cloth | |
	Japan	India	Japan	India
1893	4.28%	0.0%	5.26%	0.0%
1898	2.94	0.0	4.12	3.5
1903	5.79	0.0	7.12	3.5
1908	4.08	0.0	5.98	3.5
1913	8.33	0.0	10.97	3.5
1918	3.37	5.0	3.59	7.5
1924	1.19	5.0	3.18	11.0
1928	3.77	5.0	14.23	11.0
1933	3.02		0.81	75.0

Source: For Japan: I. Yamazawa, "Industrial Growth and Trade Policy in Pre-war Japan" *The Developing Economies* XIII-1 (March 1975): 62; for India, C. N. Vakil and M. C. Munshi, *Industrial Policy in India,* pp.41-43.

commodity markets. As Scherer (1970) put it, "what is needed for rapid technological progress is a subtle blend of competition and monopoly, with more emphasis in general on the former than the latter, and with the role of monopolistic elements diminishing when rich technological opportunities exist." These latter opportunities clearly existed, as is evident, for the case of spinning, from the correspondence between Platt and its Japanese and Indian customers. But the difference in response was related in large part to differences in organizational structure affecting the relative pressures on management to innovate.

At the time India's cotton textile industry was being established its capital markets were very imperfectly developed. The main available source of capital for early investment in the mills was consequently the personal wealth of successful merchants and financiers (Lokanathan, pp. 135-6). Most commonly, one individual would provide most of the initial capital which might be supplemented by a small group of family members and associates. This individual would then be designated as the manager of the mill, and would receive payment for this function in addition to a return on his investment. This organizational pattern, called the managing agency system, was prevalent throughout Indian industry. The managing agency would constitute a separate firm that supplied management to the mill on a commission basis, and often managed other concerns as well, including other textile mills. One special feature of the Indian cotton textile industry was the relative absence of British participation. Even though sales agents of British machine manufacturers were early participants (Koh), the industry can be characterized as having been owned and managed primarily by Indians.[13]

This managing agency system, as practised, bore several defects which undoubtedly reduced the average quality of entrepreneurial ability in the industry (Fukazawa). As noted earlier, managers were not industrialists, and they generally had no managerial or technical training or experience. They tended to concentrate on financial affairs to the neglect of the technical aspects of the mill.[14] Managerial ability was further divided between the mill, the agency firm itself, and other managed enterprises. The infusion of talent was limited by agents' long guaranteed tenure and the practice of handing down the agency from father to son. But perhaps most serious for the institutional environment was an incentive structure which often did not effectively induce the managing agent to act in the best interests of the mill. In many cases, particularly in the early period, commissions were based on physical output or sales, not on profits.[15] Even after a form of commission on profit was introduced in 1888, depreciation was not included as a cost. In many cases the managers also received, as purchasing agents, a commission on machinery purchased. The bias towards import and capital-intensity that these last two features could give is obvious. The failure to link commissions to profits, if not positively perverse in its effects, undoubtedly reduced the incentive to pursue the search for innovations intensively, especially since managing agents often sold their capital holdings in the mill at a later date.

The central defect of the Indian managing agency system was thus not that good management was prevented but that it was not adequately encouraged. Certainly there were some managers of exceptional ability. An outstanding, frequently cited, example is J.N. Tata, a prominent cotton textile mill owner who was also a pioneer in India's steel industry. In 1883, a technical advisory mission Tata had sent to England shipped two ring spinning frames to Tata for experimental use.[16] It was found that their output in low-count yarn far exceeded that of mules, and experimentation and adaptation continued under the direction of British technicians. Tata immediately began replacing mule spindles with rings, well before rings were generally accepted in Britain. Rings replaced mules in other progressive mills as well. Thus Mehta writes that Tata's lead was "followed up on an extensive scale by many others",[17] and a British report noted in 1888 that "ring throstles are very popular and several large concerns have filled their spinning rooms with these frames".[18]

The aggregate statistics on the Indian adoption of rings in Table 3, however, indicates that this practice was restricted to a relatively small number of managers in contrast to the almost instantaneous diffusion of the new, best practice, in Japan. Given substantial restraints on entry, differences in managerial and technical performance could be maintained without serious threats.[19] That most managers failed to appreciate the superiority of rings is indicated by an article appearing in a contemporary Indian trade journal, which was published in translation in the Japanese *Boren geppo* in June, 1892. The article reported:

> . . . The use of ring or so many ring spinning machines is one reason for the lack of success of the Japanese spinning industry. The Japanese industry uses many kinds of cotton and the managers of the new companies are very inexperienced. When asked why they use rings they reply that they are used with success in the United States, England and India and with such reasoning it is no wonder they have met with little success.

The same author further maintained that the short staple length of Indian cotton would continue to make the ring inappropriate. The subsequent experience of both Indian and Japanese spinning mills (large purchasers of Indian cotton) has proven this totally incorrect. The appearance of such an article in a respected Indian journal ten years after Tata successfully used rings clearly indicates that substantial portions of the Indian industry could maintain a satisfying posture—long after the Japanese had shifted entirely over to rings.

Under these conditions it is understandable that technically inexperienced Indian managers would rely heavily on British technicians; virtually all technical positions in the mills were filled by British up to about the 1880s, after which Indians began to fill only the junior posts.[20] Though it would have been impossible to run the mills at all without their contribution, this heavy emphasis on a foreign technical class probably also served to retard technological progress in the industry. First, the managers, from whose weak incentives the impetus for technological progress would have to come, found it difficult to control their technicians.[21] Second, the foreign training plus inability of the technicians to communicate with the Indian workers brought into existence another pervasive feature of Indian mill organization, the jobber, which impeded technological advance.[22]

Though the social origins of the jobber are not entirely clear, he seems to have been drawn from the same class as the workers, and so continued to function as an intermediary when Indian technicians from the middle class became more numerous. Whether through evolution or deliberate decision, jobbers were assigned middle management tasks of work supervision and labour recruitment and discipline. Once again, the incentives in operation were badly suited to induce behaviour that would profit the mill. Since jobbers could not hope to move up with the organization on the basis of the productivity or the performance of a stable, well-trained labour force[23], they concentrated their energies instead on increasing labour turnover and thus maximizing their revenue from side payments.

It is probably also true that the level of education and literacy of this middle management cadre compared unfavourably with the Japanese situation in which upward mobility and a task-oriented reward system encouraged the search for greater efficiency and technological change. In fact, how much weight to give to the differential characteristics of the industrial labour force as a whole is an interesting and unanswered question. Frequent mention is made in the literature of poor Indian work habits, such as absenteeism and long, irregular breaks.[24] The well-disciplined, hard working Japanese millhand seemed to provide a sharp contrast. Visiting Japanese missions also make much of the ennervating influence of Indian weather conditions, while Indian writers stress the low level of labour productivity associated with low levels of nutrition and health. The much lower proportion of female labour in the case of India is related to the much greater reluctance of both Muslim and Hindu women to enter the labour force. And, finally, the unwillingness to adopt double shifting is often blamed on the unreliability and individualism of the Indian worker when contrasted with his more cooperative and community-oriented Japanese counterpart. Such differences in the inherent cultural background and quality of unskilled labour might, of course, be enough to provide some bias in favour of greater capital intensity and the prolonged retention of the mule, in the case of India. But absenteeism was as high in Japanese plants and if anything, Indian operatives had more work experience (Saxonhouse, 1976, pp. 97-125). Many of the differences in performance must reflect again on a comparison of the incentives facing managers in the two industries.

Some of the deficiencies in the quality of the managerial and entrepreneurial environment in India might have been offset in part if a good social mechanism had existed for pooling technological information and experience. Such a mechanism clearly functioned extremely well in the Japanese case (*ibid*, pp. 115-118). Spinners representing over 97 per cent of the country's spindles were members of the All-Japan Cotton-Spinners' Association or *Boren*. The rapid diffusion of new technical information via *Boren* was clearly aided by its simultaneous control over the allocation of imported raw cotton to its members. In India, the organizations most closely corresponding to *Boren* were the various mill owner groups, the largest of which was the Bombay Mill Owners' Association (BMOA). Several reasons suggest themselves as to why the BMOA and its sister organizations did not serve the same technological functions as *Boren*. Their role was apparently more political, given their involvement in early controversies over tariffs, thus turning the focus away from technology. Once again this may have been a case of energies being diverted away from profits derived via more appropriate technology choices and towards profits derived from changes in the provision of government favours, credit, etc. It also seems that BMOA was patterned more after British industrial organizations; likely, without free entry, no precedent existed for extensive technology sharing. Indian managers clearly were not as effectively presented with information and advice on available technology choices as were their Japanese counterparts.

The additional technological flexibility which opens up as a consequence of quality variations usually neglected by economists undoubtedly provides substantial additional explanatory power. As we have noted earlier, a major technical advantage of rings over mules lay in the spinning of comparatively low-count yarns with the application of relatively more labour. This coarse yarn is suitable for weaving into medium or heavy weight fabrics but cannot be worked into the finer fabrics which require a higher count yarn. Before mechanization, the traditional hand crafted cotton textile products in Japan had been of the coarser type, and consumers were accustomed to cloth with these characteristics and had developed their styles around it. Further, the latitude at which Japan lies makes a warm cloth climatically appropriate. It was therefore natural for Japanese spinners to emphasize the production of low-count yarn, as was noted earlier, and then adopt rings as the best machine type. The traditions of the Indian industry before mechanization were different from the Japanese. Indian spinners had long worked in fine counts, and the greatest demand, both in foreign and domestic markets, was for fine fabrics (Chaudhuri). India's climate tends to make heavier fabrics less acceptable to consumers; moreover, more heterogeneity seems to have been demanded. When Indian textile mills began production, British imports supplied the finer yarn and cloth, leaving the lower count yarns to the Indian mills. Over time, however, the Indians moved increasingly into higher count yarns, as demonstrated in Table 8. The preference of Indian consumers for finer count yarn products, signalled through the market, was certainly one factor leading to the greater use of mules and the delayed introduction of rings in India. We have here an example of the general rule that technology choice involves not only a choice of method but also a choice of product characteristics, and that this "quality dimension" of technological choice can significantly affect the labour-intensity of the final choice. This greater "appropriateness" of indigenous Japanese consumer tastes, apparently less affected by international demonstration patterns—as well as fortuitous differences in climate favouring

Table 8   Average Count of Cotton Yarn Spun in India
(Selected Years)

Year	Average Count
1907-08	13s
1919-20	17s
1923-24	18s
1933-34	20s
1938-39	27s

Source: S.D. Mehta, *The Indian Cotton Textile Industry:
An Economic Analysis,* p. 9.

warmth over strength characteristics—thus combined to yield more labour intensive expansion paths for the Japanese textile industry.

In summary, the lessons of economic history permit us to proceed beyond the simple factor price distortions story in trying to explain comparative technological performance. Given a common source of spinning technology abroad, the choices made by the individual Japanese entrepreneur, in contrast to his Indian counterpart, illuminated the importance of differences in the institutional/organizational environment. The two sequential waves of labour using innovations adopted quickly in Japan and only slowly and reluctantly in India had less to do with the extent of overall protection than with the relative strength of workably competitive pressures at home, enhanced by the Japanese Cotton Spinners' Association structure and diluted by India's Managing Agency system. Relatively free access to credit markets and dependable technology information channels led to rapid switches of shelf technology, rapid adjustments and extraordinarily rapid rates of diffusion in Japan. In India, highly imperfect credit markets, restricted entry and a management system insulated by institutional constraints from fully harnessing entreprenurial incentives represented a situation much closer to that of the contemporary LDC. The power of machinery salesmen and multi-national corporations may be less crucial for the appropriateness of the technology choice and the direction of technology change than the ability to create and maintain an environment that places a modicum of workably competitive pressure on the industrial entrepreneur.

Notes

[1] These figures appear in Arno Pearse, *The Cotton Industry of India*, p. 22

[2] In 1882-83, the five-year average of the ratio of raw cotton imports to cotton consumption in cotton mills stood at 3.1 per cent. Raw cotton imports for this period are given in *Financial and Commercial Statistics*, Issue 4 (1897), p. 518. Cotton consumption is from Pearse, *op. cit.*, p. 22.

[3] Morris reports that mills typically worked 13 or 14 hours in summer and between 10 and 12 hours in winter.

[4] These figures agree in magnitude with an independent estimate produced by an anonymous mill manager and cited in the "Report of the Textile Factories Labour Committee" (Parliamentary Papers 1907, Cd. 3617), p. 71. He reports 28 workers per 1000 spindles, or 35.7 spindles per worker.

[5] The percentages of male, female and child labour in the Bombay mills are given in Morris, *op. cit., 66.*

[6] See Morris, *op. cit.,* p. 104.

[7] *Report of the Indian Tariff Board* (1932) p. 113.

[8] The Victorian Jubilee Technical Institute, established in 1882, included instruction in cotton mixing among its subjects. Mehta *(op. cit.)* has an illustration of a lecture on mixing given in 1896.

[9] A petition made by operatives employed in the spinning and weaving mills in Bombay, dated 24 October 1899, cited the cotton quality as a reason why Indian millhands were less productive than the English. "The breakage in the thread...is so continuous here on account of the bad quality of the cotton, that mill-owners are compelled to employ more men." See "Report of the Indian Factory Commission", Parliamentary Papers 1890-91, vol. 59, p. 107.

[10] Annual Statement of Seaborne Trade of British India, Issue 48 (1913-14), Vol. 2, p. 370, and *ibid.* Issue 64 (1929-30), Vol. 2, p. 320.

[11] A list of rates of interest on outstanding loans chargeable to the Indian government was periodically printed in the Parliamentary Papers. See also P. Banerjea, *Indian Finance in the Days of the Company,* pp. 123-5 for a discussion of interest rates in India up to the demise of the East India Company in 1858.

[12] Investment in cotton mills came from private sources, not through any institutional channel that could be regulated. Since investments branched out into many fields, entrepreneurs would not have seen an artificially low interest rate as the opportunity cost of their capital.

[13] Lokanathan (*op. cit.*, p. 136) contrasts the indigenous capital of the cotton textile industry with heavy British participation in other enterprises. Buchanan (*op. cit.*, p. 206) says "foreign capital...has had only a small place in cotton manufacture", and Mehta (*Cotton Mills of India,* p. 42) characterizes the period up to 1877 as "exclusive mill owning by Indians", though he notes an exception in 1874.

[14] See Koh, *op. cit.,* p. 127.

[15] *Report of the Indian Tariff Board,* 1927, pp. 87-8.

[16] Details are given in Mehta, *Cotton Mills of India,* p. 43, D.E. Wacha, *The Life and Life Work of J.N. Tata,* pp. 35-7, and B. Sh. Sahlatvala and K. Khosla, *James Tata,* p. 30.

[17] Mehta, *Cotton Mills of India,* p. 44.

[18]Parliamentary Papers 1888, Cmd. 5328. p. 115.

[19]See Mehta, *Cotton Mills of India,* pp. 52, 84.

[20]*Ibid.,* p. 101.

[21]*Ibid.,* p. 106

[22]See Morris, *op. cit.,* pp. 129-38, and G. Thakker, *Labour Problems of Textile Industry: A Study of the Labour Problems of the Cotton Mill Industry in Bombay,* for discussions of the jobber.

[23]An alternative explanation of high labour turnover has been given in D. Mazumdar, "Labour Supply in Early Industrialization: The Case of the Bombay Textile Industry", *Economic History Review* 26:477-96.

[24]See, for example, the comparison made between Indian and Japanese labour in Pearse, *op. cit.,* p. 11.

References

Biswas, M. R. (in co-operation with Prof. Robert Evenson, Yale University), (1980): UNCSTD in Retrospect. *Mazingira,* Vol. 4, No. 2, pp. 36-53.

Buchanan, D. H. *The Development of Capitalistic Enterprise in India,* pp 332-3.

Chaudhuri, K. N. "The Structure of Indian Textile Industry in the Seventeenth and Eighteenth Centuries." *Indian Economic and Social History Review, 11,* pp. 127-82.

Fei, J. C. H. and Ranis, G. (1965): "Innovation Intensity and Factor Bias in the Theory of Growth. *International Economic Review.*

Fukagawa. Cotton Mill Industry. In *Economic History of India 1957-1956,* V. B. Singh (Ed.), pp. 231-233.

Koh, S. J. *Stages of Industrial Development: A Comparative History of the Cotton Industry of Japan, India, and Korea.*

Lokanathan, P.S. *Industrial Organization in India,* pp. 135-6.

Mehta, S. D. *The Cotton Mills of India, 1854-1954.*

Moser, C. K. *The Cotton Textile Industry of Far-Eastern Countries,* pp. 66, 87.

Morris, M. D. *The Emergence of an Industrial Labour Force in India,* pp. 101.

Ohkawa, K. and Rosovsky, H., Eds. (1968) "Postwar Japanese Growth in Historical Perspective: A Second Look". *Economic Growth: The Japanese Experience Since the Meiji Era.* Richard D. Irwin, Inc.

Ranis, Gustav (1959) "The Financing of the Japanese Economic Development". *Economic History Review XI,* April, pp. 440-454.

Saxonhouse, G. (1976) "Country Girls and Communication Among Competitors in the Japanese Cotton-Spinning Industry", in H. Patrick, ed., *Japanese Industrialization and its Social Consequences,* pp. 97-125, 115-118.

Sarkar. *The Economics of British India,* pp. 216-7, 221.

Scherer, F. M. (1970): *Industrial Market Structure and Economic Performance,* pp. 378. Rand McNally College Publishing Co., Chicago.

CHAPTER 2

Technology Choice and Technological Change in Third World Agriculture:

Concepts, Empirical Observations and Research Issues

Carl H. Gotsch
Stanford University

and

Norman B. McEachron
SRI International

INTRODUCTION

THAT TECHNOLOGICAL CHOICES both affect and are affected by a wide variety of economic, social and political institutions is by no means a new idea. It was clearly a part of the world view of the classical economists, most notably Karl Marx, and more recently has captured the attention of groups as diverse as modern econometricians and Nader's Raiders.[1] At one point, social scientists were content to focus largely on the impact of technical achievements on the existing organizational and social structure; much of their current attention, however, is being devoted to the question of how attitudes, values, and institutional arrangements act to determine the sorts of technology-development activities that are undertaken in the first place. The problem of technology choice is thus seen, not simply as one of choosing from among a set of feasible alternatives at a particular point in time, but as one in which the size and composition of the feasible set is itself an issue.

The following paper deals with the interaction of technology and the surrounding socioeconomic structure at several different levels. First, an effort

29

is made to develop a broad conceptual framework in which various pieces of the problem can be located. The basic ideas draw heavily on Ruttan (1978) and de Janvry (1970) but have been extended to include our own interest in the private sector and the linkage between industrialized and developing countries.

The second section relates the conceptual "system" to a detailed discussion of technology choice and technological change in Pakistan agriculture. Particular emphasis is given to an analysis of the sequence of induced mechanical innovations and the efficiency with which demands from the farmer's point of view were translated into supplies of improved inputs.

Lastly, the paper juxtaposes the conceptual framework of the first section with the data available in Pakistan as a basis for suggesting some research priorities. It is altogether evident that much more is known about how technology influences its economic and social environment than how these factors determine the direction of R&D expenditures. Relatively little detailed empirical work has been done, for example, on how and why research resources are allocated at the micro level. Studies have established the importance of relative factor scarcities in determining long-term trends in overall resource commitments. However, in the short-run, casual observations suggest that a much wider spectrum of considerations are at work, both in allocating resources to agricultural research and, within agriculture, to regions, commodities and social groups.

A SYSTEMS FRAMEWORK FOR INVESTIGATING TECHNOLOGY CHOICE

EVEN A CURSORY PERUSAL of the multiple, often interweaving strands of thought that characterize the recent literature on the generation and implementation of new technology suggests the desirability of adopting a more explicit "systems" view of the process. Such a framework would consist of:

1. A series of major categories of systems elements, or *variables*.
2. A series of *feedforward and feedback loops* linking the categories that represent the fundamentally dynamic character of the technology development and choice process.

The systems framework attempts to add additional coherence to the myriad factors affecting technology generation and to identify areas in which understanding is particularly weak or the potential for policy intervention particularly good.

Although a comprehensive systems formulation is admittedly open to the charge that the pervasiveness of technology in society may make attempts to trace its ramifications too unwieldy to be helpful, the opposite danger of seeing this particular problem in a partial sense appears to be more threatening. For information on any one aspect of the overall picture to be fully appreciated, some systematic effort to set out the dynamics of the entire process is a necessary precursor to more conventional and specific examination of its various parts.

The origin of the framework used below can be traced most directly to de Janvry's 1978 formulation, but the approach is in keeping with the views of other authors who have suggested that the determinants of technological choice include, along with engineering and economic considerations, bureaucratic and political motivations as well.[2]

An adaptation of de Janvry's work is given schematic representation in Figure 1. The most obvious point of departure in using it to organize information is to assess the economic benefits to firms of obtaining additional technological alternatives. This assessment takes the form of a "benefits" matrix, one dimension of which is a set of resource constraints that could be relaxed by employing a new technology and the other dimension of which is a sorting of firms by a characteristic (e.g. firm size or magnitude of output) that affects the value or economic benefit to the firm of obtaining additional alternatives. Each entry in the matrix equals the economic benefit, (e.g. the increased profits) of relaxing a particular constraint by a unit amount for a firm with a particular set of characteristics.

In keeping with the induced innovations argument, the benefits to relaxing the resource constraints (in a programming formulation of the model) are assumed to point toward technology that would: (1) augment scarce resources; or (2) represent an alteration of the technical parameters for activities that are profitable but constrained by limited resources.

The second step in organizing material in a systems framework involves an analysis of the bureaucratic and market responses to the "signals" emanating from commodity producers. In an open economy with transnational interactions, three private, market-oriented channels of response exist, albeit with varying levels of government regulation: (1) domestic private-sector R&D and technology adaptation; (2) foreign direct investment, licensing, and other forms of international technology transfer to the private sector; and (3) private imports.

Three similar responses can be identified in the public sector. However, in the public-sector case, the analysis must be broadened to include observations on the relative effectiveness of different social groups in translating latent demand into the allocation of resources to particular research problems.

The third step required to round out the systems framework and to make it dynamic is to re-introduce the supply of technology actually produced into the previously constructed farming systems model. Comparison of the model results with predicted demands permits a further sharpening of hypotheses about the determinants of technological choices as seen in practice. This third step also becomes the basis for policy prescriptions about the types of activities that need to be adjusted in the system if the mechanisms that translate latent demand into a supply of technology are to reflect exogenous social values, such as a more equitable distribution of income or increased employment opportunities.

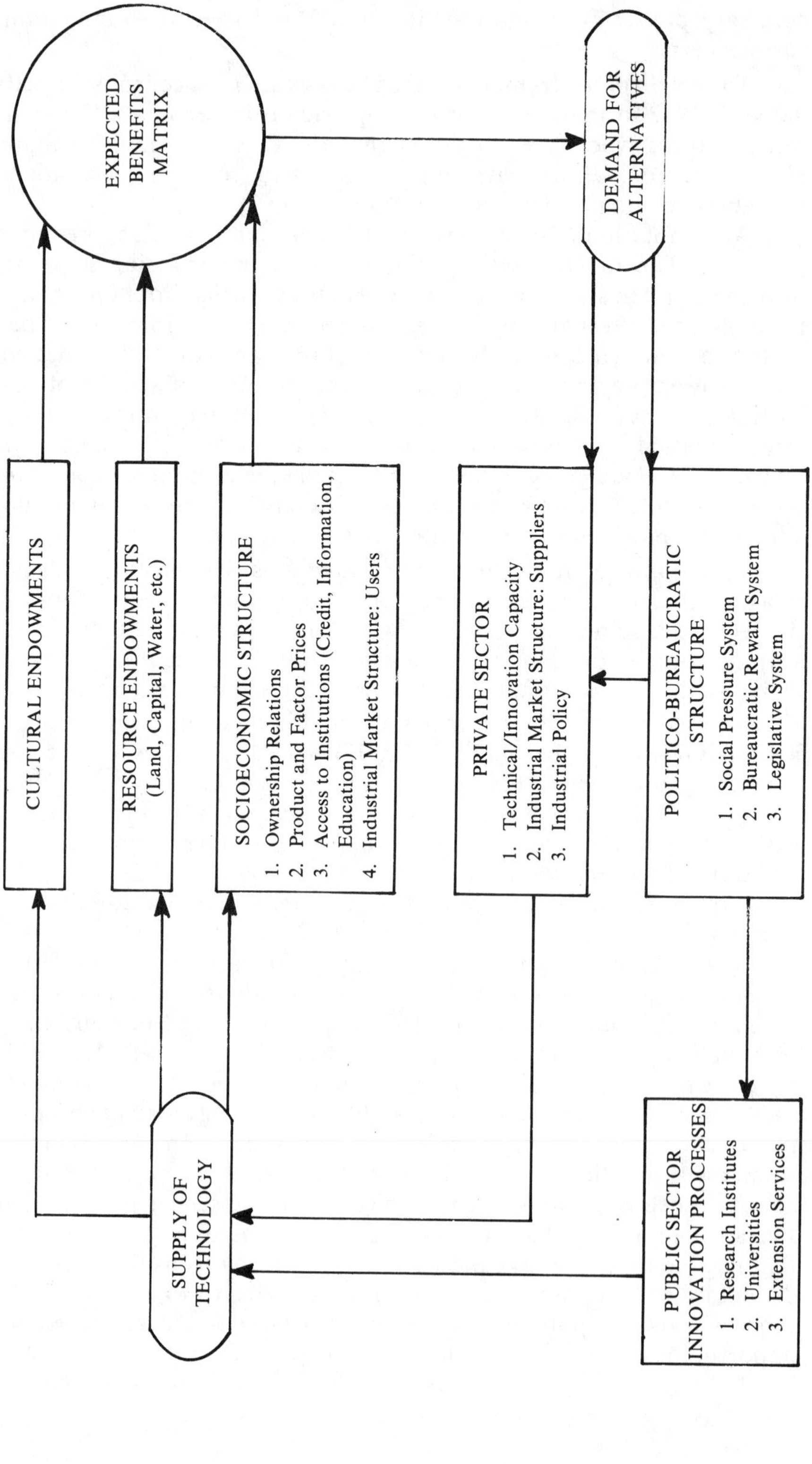

Figure 1 Supply and Demand for Technological Alternatives in a Closed System

The focus throughout the framework and its subsequent applications is on the micro level. Macro policies naturally come into play as government price and exchange rate policies, credit rationing, tenure relationships, etc., have an impact on the normative and positive models. However, the level of investigation, whether it be the construction of a representative farm model or the presentation of surveys on light industry, is on the task-specific nature of technology. Reasonable men may, of course, differ on the level of generality that is the most appropriate for examining the problem of technological choice; but the perspective of this study is the machine, the labourer, and the specific good that is to be produced.

DETAILS OF A CLOSED ECONOMIC SYSTEM

FIGURE 1 IS LIMITED to the variables and linkages for technology supply and demand in a closed economy (i.e. one having no economic transactions across its national boundaries). The framework, as noted above, incorporates social, political and institutional actors, as well as those economic and engineering considerations that have traditionally been the subjects of investigations of technological change. Thus the framework accommodates both the "radical" view of technological change which stresses the role of social, political, and institutional elements in determining the pattern of choice, innovation, and diffusion, as well as the more neo-classical view, which sees the wellspring of change in the emergence of relative factor scarcities.

The Benefits or Demand Matrix

The starting point in developing empirical materials relevant to this formulation is the *ex ante* "benefits matrix". This matrix specifies the net gains and losses that any particular group selecting a technology expects to receive from the implementation of any combination of technical alternatives. It can also be broadened to include institutional alternatives that affect the benefits and costs of a technology (e.g. hire-service markets, access to credit and information).

The concrete expression of the benefits matrix in the subsequent empirical investigation is a synthetic micro model—a "representative firm"—that incorporates factor scarcities, produce and factor prices, and technical parameters of alternative production processes now in existence. The representative-firm model predicts the choice of technology (e.g. the choice that maximizes profits), accounting for technical and economic elements and for aspects of risk aversion and institutional structure that are deemed important by alternative formulations of the constraint set.

Solutions to the optimal allocation of resources for representative firms provide not only estimates of the optimal decision variables, but the economic value to the firm's proprietor of: (1) augmenting scarce resources; or (2) altering the technical parameters of activities that are profitable but constrained by scarce factors.[3] These scarcity values constitute incentives that a firm's proprietor would have to seek technologies not currently in the available choice set, i.e. the incentives (or latent demand) for technological innovation.

Empirically, the sources of information utilized by decision-makers in searching for improved technology is an important variable worth investigating. Information sources will include not only publications and official channels, but, even more important, individuals in the decision-makers' personal communication networks. Research on the diffusion of innovations has shown that interpersonal communication is relatively more important in an innovation adoption decision than are media (Rogers with Shoemaker, 1971). Since interpersonal networks are largely determined by socioeconomic background and education, it would be natural to hypothesize that educational training and ties to an international scientific and technical community heavily influence technology choice.

Technology Supply Response

The determinants of the demand for technology and for the emergence of institutions such as hire-service markets by private-sector producers flow directly from the assumption that firms seek profits subject to the constraint of resource scarcities. Much less well understood are the mechanisms by which these latent demands are transmitted and acted on by those who supply technology. The second step in adding empirical content to the systems framework is therefore to examine the effects of the economic inducements for technological and institutional alternatives on the existing channels of response.

In a closed economy, there are two such broad supply channels. The first is market-oriented and depends on private initiatives in R&D, engineering, and technology adaptation and marketing. The second channel is implemented through the intervention of public-sector organizations in the form of government-performed and/or government-sponsored R&D, engineering, and diffusion efforts.

Technology Supplied From the Private Sector. The more prominent of the two channels through which demands for and supply of technology flow, at least in the industrialized countries, is the response of private input suppliers as they too seek profits in the marketplace. Through constant monitoring and interaction with their clients, they anticipate those kinds of technical change that will save labour, improve productivity and output—and, on occasion— create indirect forms of social control. Although perhaps not as well-developed in the Third World, the private sector has also become an important supplier of technology in a number of developing countries. The light-industry sector that has emerged in India and Pakistan, for example, has played an important role not only in supplying machines to the agricultural sector but in adapting these machines for the needs of the small farmer community. Unfortunately, as another section of the present study indicates, little is known about merchandising, promotion, and other such techniques that constitute the links between demanders and suppliers.

Technology Supplied From the Public Sector. In general, the determinants of technology that emerge from the public-sector channel involve a much more complicated set of objectives as compared to private business. For example, the ability of different social and political groups to make their influence felt is an important ingredient in the way research resources are allocated.[4] The obvious difference in the ability of social groups to affect the role of public sector organizations has made these institutions as controversial in agriculture as that of "agri-business". Considerable evidence exists, for example, that in most countries, politicians and bureaucrats have considered research suggestions by large farmers more favourably than the needs of small farmers or landless labourers. In part, this is simply a reality of political life. The larger farmers are the supporters of the politicians in power and, as a direct consequence, inevitably have an influence on the researchers (1) who work on government lands; or (2) who receive substantial amounts of government research funds. Less support for small-scale industry than for large, capital-intensive facilities can be understood in similar terms.

Detailed examinations of the indigenous scientific and technical community—which includes the universities and R&D—underscore general impressions about public sector performance. Unfortunately, these organizations often play a critical role in the diffusion of technology for they are often the conduit between the international community and local institutions. Ideally, they would serve as a screen to select or adapt technology that was most suited to local needs. With few exceptions, however, the performance of indigenous R&D facilities in developing countries has not been spectacular.

A major source of the difficulty seems to be an inadequate linkage to indigenous users (Sunkel, 1971; and Vietorisz, 1973). As Sunkel notes:

> For a techno-scientific literature to influence the process of development there must be close institutional links with the productive structures of society as well as with the political and administrative structures that control the decisions and the field of development policy. The lack of these links, rather than the actual amount of research, is the main characteristic of technical and scientific underdevelopment.

Much of the difficulty can be traced to a larger problem stemming from the organization of indigenous R&D. In many developing countries: (1) practically all R&D is funded by the government; and (2) most R&D is conducted in government and university laboratories. In far too many cases, the result is that potential industry users are not involved in the R&D process. Channels of communication between the R&D establishment and users are typically minimal. It is therefore not surprising that a field researcher who visited 50 research centres in 13 developing countries concluded that, in general, their activities were not relevant to domestic problems (Blackledge, 1972). Domestic firms apparently did not have confidence in the results that the research centres could produce, and the centres were not aware of the needs of potential users. This also appears to be true of research in universities in developing countries, which tends to be oriented toward the international scientific community. Local problems, and specifically the problems associated with social and economic development, receive little attention (Herrera, 1973).

The few research centres of quality that do exist in developing countries are often closely linked to the scientific systems of developed countries. They function as isolated enclaves that have only minimal concern for the development of technology that responds to broadly based local needs.[5] However, one of the very few in-depth studies of scientific communities in developing countries (McCarthy, 1972) discovered that scientists located at regional universities in the Philippines were more likely to engage in research with a Philippine focus. These scientists were relatively isolated from the international scientific community and instead had contact with industry, large agriculturalists, and professional and community groups. Although there was no attempt to evaluate the relative success of the different universities, the study concluded that the presence of a "mutual delivery system" linking scientists to other segments of the society had positive attributes.

Organizational issues have also seemed to affect the Indian scientific establishment, especially in the industrial area. A report of the Government of India concluded that India's chain of national laboratories had failed to produce a significant impact on the country's technological progress (Government of India, 1970). Very little of the research output was of use to industry. The report also pointed out that there was no arrangement for enabling scientists to visit industrial establishments and acquire a detailed knowledge of industry problems.[6]

The Socio-Economic Screen

In the next step of the conceptual systems framework, the supply of technology and institutional alternatives that emerge between time *t-1* and time *t* are filtered through the set of existing economic and local institutional conditions (e.g. tenure relationships, product and factor prices, market structures, access to information, etc.) to produce the predicted impacts on the industrial or agricultural community in a particular locale.

A major structural determinant of the benefits of adoption is the relationship between technology and firm size or output. A key to this relationship is the divisibility (or economies of scale) of the technology. For technologies that are divisible (or have little or no economies of scale), firm size is not a long-run barrier to adoption (e.g. improved seeds and fertilizer in agriculture). On the other hand, a highly indivisible technology, or one with substantial economies of scale, favours larger firm size. If the technology is such that a timestream of services cannot be provided to firms of small size, as is typically the case with industrial technologies and with certain fixed-location agricultural technologies such as water supply, then hire-service markets or other institutional innovations may not make their appearance to alleviate the problem (i.e. to reduce the extent to which the technology favours larger firms). One result can be the reduced economic viability of small-scale and more labour-intensive production units in favour of larger-scale and more capital-intensive units.

Although debate continues on this topic, there is now considerable evidence that, in agriculture, if the innovation is highly divisible (e.g. improved seeds and fertilizers) firm size is not a barrier to adoption. Sometimes, there is a lag of several years, but even in the face of credit constraints, the actual benefits matrix that describes what various social groups have achieved bears a close resemblance to the *ex ante* matrix that describes what they could achieve.

The evidence with respect to mechanical technology in agriculture or industry, however, is not as encouraging as the evidence on biological and chemical technologies in agriculture, because indivisibility is common and not always subject to the development of hire-service markets. For example, whereas their mobility and alternative use as transportation vehicles have caused tractor service markets to develop rather quickly, the markets for water—a crucial production input in arid and semi-arid agricultural areas—are much less developed. Tubewells, low-lift pumps and motors cannot be easily moved, and consequently the hire-service arrangements that transform a stock of technology into a flow of services are often slow to make their appearance. Similarly, much industrial technology is immobile and not subject to hire-service arrangements, although sharing capacities of machine tools among alternative jobs, for example, serves a similar function.

Completing the principal feedback loop in the systems model involves reintroducing technology actually produced back into the previously constructed representative-firm model. It is to be expected that a comparison of the *ex-post* benefits matrix with what might have been predicted *ex-ante* will show substantial differences. In part, these differences are a function of the technology that has been made available. They are also heavily dependent, however, on the decision to adopt on the part of potential users. Particular attention in this regard needs to be paid to the availability of information about alternative technologies and the characteristics of the decision-makers who select among these technologies—particularly in industry, where the choice of technology and the interdependence among production processes is, in many instances, exceedingly complex. The representative-firm model, as well as most research analyzing the choices of technology in developing countries, is conducted under the assumption that entrepreneurs and farmers are rational, well-informed profit and utility maximizers. This is clearly an idealization. As Stewart (1977) points out:

> Those who introduce techniques into underdeveloped countries thus make a choice among the techniques available; the choice actually made depends on the nature of the decision-makers and their objectives, the economic circumstances in the economy concerned, and the characteristics associated with the different techniques, bearing in mind that their choice is confined to techniques they know about, and that knowledge may often be incomplete or inaccurate.

A second alternative that may account for differences between what is seen as "potential" from a technical and economic point of view rests with the notion of profit maximization as a goal. Ranis (1978) notes:

> A basic limitation of the traditional approach which became increasingly apparent during the course of our research is the identification of the entrepreneur as a rational profit maximizer concerned entirely with the economic environment and the economic calculus within that environment. ... It is the understanding of decision-making which goes beyond the short-term profit maximizing calculus which is likely to illuminate what actually transpires.

A variety of other studies have pointed to both problems of limited information and "bounded rationality" as explanations of why potential innovations are not taken up as rapidly as might be expected or, for that matter, not taken up at all. On the other hand, examples also abound in which lower production costs and increased profits provide an incentive for engaging in a more active search process.

THE SYSTEMS FRAMEWORK IN AN OPEN ECONOMY WITH TRANSNATIONAL ACTORS

THE SYSTEM DESCRIBED ABOVE assumes that the supply of technology is generated by the response of domestic institutions to a local demand for technological innovations. Were it not for outside influences, it would be expected that over time, the forces of supply and demand would produce a rough kind of equilibrium that would constitute one element of what has come to be called the "structural transformation" of an economy. However, many of the major issues surrounding the problem of technological choice in developing countries stem from the interactions of their economies with the very different economies of the industrialized countries across national borders. Figure 2 attempts to capture these relationships by adding a second system to the picture and indicating several feedback mechanisms that have become primary sources of concern for planners interested in problems of income distribution and employment. Of particular significance in the diagram are the "transnational actors" that have been interposed between the two domestic systems. Much of what happens in the area of technological transfer is the result of the perceptions, modes of operation and sources of

Figure 2   An Open System with Transnational Interactions

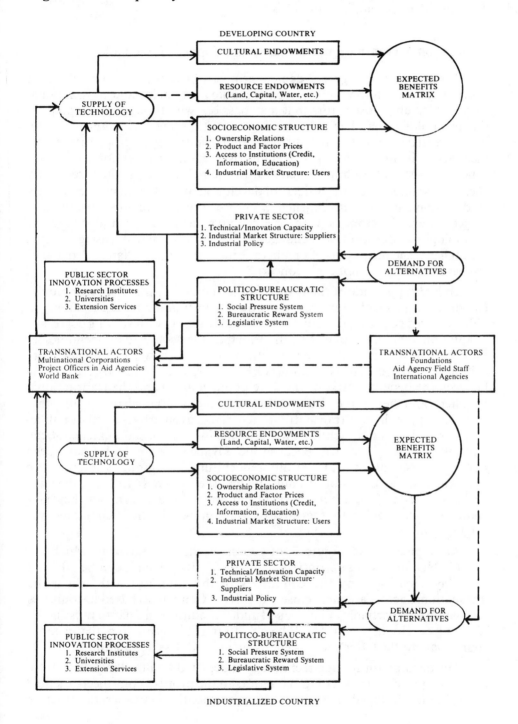

information of organizations that are interposed as filters between the developed and developing countries.

Conventional Transnational Demand and Supply of Technology

The left side of Figure 2 depicts groups whose activities have been of greatest concern in the literature on technological choice and transnational actors.[7] Namely, those actors that may pass on, with little or no adaptation, the technology that has been generated within the environment of an industrialized country through direct foreign investment. In the private sector, these actors include transnational enterprises, equipment vendors, and other export-import and trading companies. In the public sector, they frequently include, as stereotypes, the officers of development banks and project officers of development assistance agencies. The relationship of these profit- and project-oriented actors is with the effective demand for and supply of technologies and institutional changes in the various countries, as aggregated through market and politico-bureaucratic institutions. The "latent demand of the benefits matrix is not their concern".[8]

On the supply side of the system, these actors may transfer technologies embodied in products, services, blueprints, and training from the supply in industrialized countries directly to the available supply with varying degrees of adaptation, including none at all. In principle, they could adapt technologies no longer in use in their home countries or invent new technologies more suitable for the developing country. However, there is no direct linkage between the benefit matrix of the developing country and the transnational suppliers and hence the demand for technology (and virtually all other relationships with these transnational actors) is screened by market or public-sector institutions representing the *status quo* in terms of economic, political, and bureaucratic influence. Because the advanced technologies have been developed in countries with very different cultural backgrounds and resource endowments, the presumptions must be that they will not "fit", i.e., they will be inappropriate for all but the small minority of the enterprises that have characteristics similar to the environment in which the technology was originally developed.

An implication of Figure 2, of course, is that the extent to which the original technology is appropriate or the ease with which it can be adapted varies systematically by the conditions in the countries with which the transnational actors are most closely involved. Companies based in countries with vintages of technology developed under conditions similar to those in the receiving developing country will be able to transfer technology that is more appropriate to the latter's factor endowments with less adaptation.

The demand for foreign technology is depicted as a feedback loop from less developed countries' (LDCs) interests to the profit- and project-oriented actors in the developed areas. In this loop, the politico-bureaucratic structure of the developing country regulates both international relations and domestic institutional policies such as relative factor prices, access to credit, etc. Strengthening of this regulatory function to reflect the interests of the developing countries has been a principal concern of LDCs in recent years, as articulated at UNCTAD and the UNCSTD meeting. As Figure 2 suggests, however, effective strengthening of the interests of the poor majority in the policies pursued by governments of developing countries cannot really occur until institutional reforms permit these groups to have a greater voice in their own societies.

An Emergent Transnational Linkage of Demand

The second transnational feedback mechanism in Figure 2 links the demands generated by the benefits matrix in the developing country to the demands being made in the industrialized country. It has been depicted as a dotted arrow because of its recent emergence and the uncertainty surrounding its significance. Perhaps the most visible testimony to its presence is contained in recent legislation in the U.S. Congress directing the Agency for International Development to concentrate on the problems of the "poor majority" in developing countries and to move away from large-scale, capital-intensive infrastructure projects that tend to benefit most those who already are participating substantially in the economic arena.

Other transnational groups have also attempted to create at least a latent demand for a more direct response to the benefits matrix generated by the developing country's resource endowments and socio-economic environment. Among the action agencies, for example, certain staff groups within the World Bank have begun to try to identify ways in which the technology choices made in the traditional project departments would reflect, not only economy-wide scarcities but micro-level concerns of employment and income distribution as well. Lastly, philanthropic and educational institutions active in development have attempted, through conferences, seminars, and studies, to argue that very often the signals generated by the benefits matrix are seriously distorted in the host countries themselves and that this process is too often abetted by the direct market and bureaucratic linkages between the developed and developing countries.

As the dotted arrow in Figure 2 is intended to suggest, the demand created by transnational actors who are not actually involved in the production and sale of technology must first be realized through decisions made in the marketplace or by bureaucratic machinery of the more developed country. Moreover, the figure also suggests that this process, avoiding as it does the domestic political and bureaucratic structure, will encounter resistance if made a condition of the supply of technology. Obviously where the domestic political and bureaucratic processes in the developed country are already attempting to respond to the demands expressed in the benefits matrix, the divergences of views will not be important. But where this is not the case, the experience of the past few years indicates that efforts by developed countries to make a concern for income distribution a criterion for the transfer or funding of technology may be met with a good deal of indifference (Dunkerley, 1979).

Additional general observations could be made about the systems depicted in Figures 1 and 2 and the role of information flows and information processing in these systems, particularly in the form of anecdotes that make concrete the linkages to which attention has been drawn. However, a systematic exposition of particular situations in which as many elements as possible were drawn together is probably more instructive. Attempting to develop a holistic view of the problem, especially in terms of its internal dynamics, is no substitute for the detailed, link-by-link investigation of the hypothesized relationships. Indeed, it is in these specific causal relationships that the requirements of scientific evidence can be most rigorously applied and tested.

IMPROVED MECHANICAL TECHNOLOGY IN PAKISTAN AGRICULTURE

THE SYSTEMS FRAMEWORK described in the preceding section is useful in providing an overview of the technology choice process. It also assists in organizing a research effort in such a way that the various parts of the investigation add up to a more interesting whole. But it does not indicate the particular research methodology that should be taken at each point in the project. Indeed, given the multi-disciplinary characteristics of the systems framework, one rightly suspects that a relatively eclectic approach will be required. As noted earlier, however, a sensible point of departure in understanding a particular industry seems to be the construction of a synthetic, micro model—a "representative-firm". In addition to its role in systematically organizing data on potential activities or processes, such a model can also be manipulated to simulate a variety of policy decisions and institutional situations. Hypotheses generated by the task-specific, disaggregated model can then, in the second phase, be subjected to a more direct cross-sectional statistical examination.

The second step of the analysis (i.e. the examination of the determinants of technological choice from the supply side) often requires other, less formal techniques. This is particularly true when the area of interest involves bureaucratic or political response to technological demands or needs. The number of senior policymakers in such situations is ordinarily limited, and their attitudes and perceptions of the problem are better dealt with in the descriptive mode of political science or public administration. Cross-sectional survey methods, on the other hand, can be used to ascertain the effectiveness of markets in transmitting signals to small-scale manufacturers for the development of various types of machines and equipment.

The third step of the analysis (i.e. filtering of a newly available supply of technological and institutional alternatives through the set of existing economic and institutional conditions to produce a predicted set of impacts on the industry) involves the introduction of the alternatives that are actually observed into the representative-firm model. This step yields a prediction of the technologies that will be adopted. Such adoption leads to a relaxation of some resource constraints, to a new set of scarce resources and associated shadow prices, and thus to a new set of incentive signals for the supply of additional technology to address. This third step, then, becomes the first step of a new round of analysis.

This analytical framework is applied below to the generation and choice of mechanical technology in the Pakistan Punjab. The mechanical technologies involved are groundwater supply technology (in the first round of analysis) and tractors and threshers (in the second round). Each round of analysis is organized according to the three steps described above. In terms of the systems framework, it is a relatively simple case, but one that permits an uncluttered exposition of the conceptual approach. In particular, it (1) calls attention to the potential interaction between the private and public sectors in technology generation and diffusion, and (2) emphasizes the dynamic character of technological choices in general. The subsection that then follows takes up the more complicated determinants of the diffusion of tractors and threshers as part of a second round of technological change.

GROUNDWATER DEVELOPMENT IN THE PAKISTAN PUNJAB

Step I: Benefits Matrix for Traditional Agriculture

SEVERAL LINEAR programming models already exist from which the impacts of different types of technology on the scarcity value of domestic resources can be ascertained.[9] Table 1, for example, shows the scarcity values imputed to the various resources and special constraints in the production process without the availability of the traditional Persian wheel as a source of supplementary water supplies.[10] As noted previously, these scarcity values or shadow prices are the key to the induced-innovation hypothesis. Not unexpectedly, without a Persian wheel, the scarcity value of water dominates the determination of the cropping pattern in the model. The shadow price is highest for September, when summer crops are maturing and preparatory tillage is underway for the fall crops.

Punjabi farmers, by digging thousands of shallow wells, contributed significantly to total irrigation requirements. Table 1 also shows that even the

Table 1　Scarcity Values of the Binding Constraints for a 12.5-acre Farm With and Without a Persian Wheel (Rupees per Unit of Resources)

Resource Constraints*	Scarcity Value	
	Without Wheel	With Wheel
Water (acre-inches)		
April		.73
May		3.44
June	3.71	.73
July	7.59	.73
August	16.81	.73
September	59.94	.73
October		12.21
November		18.44
December		.73
February		1.13
Land (acres)		
October		73.31
Bullock power (hours)		
May		.42
October		1.26
November		1.95
Labour (hours)		
April	.30	
May	.40	.40

* Months for which no resource constraints are binding have been omitted.

small amount of water produced by these primitive devices can alter the entire set of scarcity values. Because animal power has a low opportunity cost in the late summer months, it can be used extensively for water production. The result is a change in the model's optimal cropping pattern such that surplus water from surface sources can be used productively.

According to the model, the possibility of producing additional water with a well and Persian wheel also made land and animal power scarce resources. This was certainly to be expected in the case of the latter; the scarcity value of animals is now determined largely by the scarcity value of the water they produce. However, the additional water also creates a situation in which crops compete for land in October. The implication of this result is that where shallow wells can be used to reach groundwater, additional power, or crop varieties with shorter growing seasons, would be potentially interesting innovations.

Step II: Induced Bureaucratic and Market Responses

Based on the benefits matrix that includes the shadow prices shown in Table 1, the induced-innovation argument would suggest that the technology-supply response would be the provision of water-producing devices. This is indeed what happened. In the early 1960s, both the private and public sectors rapidly became involved in the provision of pumps, motors and tubewells for areas in the Punjab that could exploit the area's groundwater potential.

It should perhaps come as no surprise that, given the potential payoff shown by the shadow prices in Table 1, there were both private and public responses. Although each was being paid in a different coin (i.e. small-scale manufacturers were interested in money while bureaucrats were interested in the political benefits) each did recognize the signals represented in the benefits matrix and made an effort to respond.[11] It appears that machine-tool and foundry skills cultivated in this particular region for decades were used to create an industry that exported water-producing machinery over a radius of several hundred miles.

The initial response of the public sector to the potential for groundwater development was, to a certain extent, the by-product of efforts to deal with what was regarded as a more pressing problem, namely, the increasing incidence of waterlogging and salinity. The proposed solution to the problem was to use large tubewells to provide "vertical" drainage instead of the more conventional methods of laying tiles and collecting excess water in field drains. Subsequently, it became apparent, at least in the fresh-groundwater areas, that lowering of the water table not only would provide drainage but would add significantly to available irrigation supplies as well.[12] Unfortunately, however, experience with the public sector programmes also revealed that the managerial requirements of administering such large, integrated schemes were, at least in the early stages, beyond the capacity of local organizations. Whereas the demonstration effects of the SCARP programme with respect to the value of groundwater development belongs on the credit side of the ledger, the programme's demonstration of mismanagement, inability to repair pump breakdowns, electricity failures, etc., provided a reminder of the limited institutional capacity available in the country.

Step III: Socio-economic Structure and Adoption

The initial benefits matrix established the desirability of augmenting traditional supplies of irrigation water, and the market and the government bureaucracy both responded. What has been the result of diffusion process, particularly in terms of the different matrices that distinguish various social groups in the rural areas?

The difficulties encountered by the government in managing the public-sector programme have been mentioned above. These were sufficiently severe so that, when large-scale foreign assistance disappeared with the India-

Pakistan war in 1966, progress on the programme virtually came to a halt. Only recently has it been renewed under World Bank auspices.

Whatever may be said about the inability to manage the SCARP

Table 2 Optimal Cropping Patterns and Cropping Intensities for a 12.5-acre Farm in the Central Punjab Wheat-Cotton Area under Alternative Technological Assumptions (Per cent except as otherwise indicated)

Crops	Traditional Technology		Improved Wheat and Rice Technology	
	Without Tubewell	With Tubewell	Without Tubewell	With Tubewell
Winter crops				
Wheat	43.5	8.8	44.3	34.3
Barley				
Oilseeds	2.6			
Gram				
Fodder	14.0	12.0	6.9	5.0
Sugar cane		6.9	3.7	5.6
Vegetables	1.1	1.4	1.7	1.1
Orchards		2.1		
Subtotal	61.2	31.2	56.6	46.0
Summer crops				
Rice	2.3		7.7	37.7
Cotton	25.5	50.8	20.1	
Maize				
Fodder	10.2	9.2	10.4	7.5
Sugar cane		6.9	3.7	5.6
Vegetables	.8	.6	.8	.5
Orchards		2.1		
Subtotal	38.8	69.6	42.7	51.3
Total	100.0	100.0	100.0	100.0
Total cropped acreage (acres)	11.8	14.5	11.5	17.7
Cropping intensities	94.0	116.0	92.0	142.0
Net revenue (rupees)	2,034	3,192	2,475	4,131

Table 3 Scarcity Values for the Optimal Cropping Patterns in the Wheat-Cotton Area (Rupees per Unit of Resource)

Resource/Crop Constraints	Traditional Technology		Improved Wheat and Rice Tech.	
	Without Tubewell	With Tubewell	Without Tubewell	With Tubewell
Water (acre-inches)				
April		1.00		1.00
May		1.00		1.00
June		1.00		1.00
July		1.00		1.00
August		1.00	12.00	1.00
September	17.20	1.00	14.30	1.00
October	29.70	1.00	2.60	1.00
November		1.00		1.00
December		1.00		1.00
January		1.00		1.00
February			1.00	
March	.10		10.20	1.00
Land (acres)				
September		16.60		58.80
October		159.00		196.20
Labour (hours)				
April	.20			.40
May	.40	.40		.40
June			.40	.40
September				.40
October		.40		.40
November		.40		
December		.40		

programme, however, its distributive effects at the micro-level were generally beneficial. Siting large tubewells at the head of the watercourse meant that everyone, regardless of farm size, received a larger volume of water than previously. This proved particularly beneficial to the smaller farmers who could now, with what had formerly been excess animal and manpower, farm holdings more intensively.

A rather different picture emerged from water-producing activities in the private sector. Table 2 shows the dramatic impact of tubewells on resource allocation in the representative-farm model while Table 3 shows the impacts on the scarcity values of the resource constraints.[13] It also suggests that the impact of increased water supplies accentuated the effects of the high-yield varieties (HYV) technology that became available several years later. Some aspects of Table 2 are predictable (e.g. all months are sufficiently short of water so that water-producing activities enter the feasible solution at some level every month). Consequently, the shadow price of water shown in Table 1 for the representative farm declines significantly in Table 3 to simply the cost of pumping water. More important in terms of signals for the next round of technological change are the new shadow prices for land, labour and animal power.

The representative farm from which the model's parameters were derived was assumed to have a family and a pair of bullocks for power. The tubewell was assumed to be jointly owned or provided by the landlord for a group of tenants. As Table 4 suggests, the impact on the respresentative farm is very different if various types of indivisible mechanical technology are required to enter the optimal technology package in discrete (integer) units. According to model results, in 1959-60, farmers having less than 15 acres did not find it

Table 4   Technology Packages for Different Sizes of Representative Farms: Integer and Continuous Solutions

Resources	Farm Size (acres)							
	5	10	15	25	35	50	75	100
Mixed integer solution								
Bullocks	0	2	2	0	1	2	1	2
Persian wheel	0	1	0	0	0	0	0	0
Stationary engine	0	0	1	1	1	1	1	1
Tractor	0	0	0	1	1	1	2	3
Tubewell	0	0	1	1	1	1	1	1
Thresher	0	0	1	0	0	0	1	0
Continuous solution								
Bullocks	.17	.33	.50	.83	1.17	1.67	1.71	1.40
Persian wheel	.00	.00	.00	.00	.00	.00	.00	.00
Stationary engine	.03	.07	.10	.17	.24	.35	.52	.66
Tractor	.12	.24	.37	.61	.85	1.22	1.96	2.75
Tubewell	.03	.07	.10	.17	.24	.35	.52	.66
Thresher	.03	.07	.10	.17	.24	.35	.50	.66

Note: Integer entries are numbers of each item predicted to be used; decimals are fractional usage rates.

desirable to invest in the 1-cubic-foot-per-second wells that were the standard output of the Daska machine shops. Consequently, they were unable to break the water constraints that traditionally limited agricultural intensity in the Punjab and hence were unable to utilize their excess labour resources more effectively. The effect, at least in terms of the model results, was greater income disparity between farm sizes, even in the presence of a technology that should have been the key to a more labour-intensive agriculture.[14]

Two questions occur at this point. Given the shadow price of water, did hire-service markets emerge that helped to create a flow of resources from the "lumpy" stock represented by the machine itself? Alternatively, was there an effort to redesign the technology in such a way that it was more compatible with the existing distribution of farm sizes? The answer to the first question is a qualified yes. Markets have arisen in water, although the stationary characteristics of well installations obviously limit the potential for farm-to-farm water transfers. Cross-sectional farm management surveys confirm this intuitive result. One of the most carefully done studies shows, for example, that although there appears to be very little difference between small and large farmers in their adoption of such divisible technology as fertilizers and improved seeds, there is a distinct difference in the amount of supplementary irrigation water used.[15]

The answer to the second question appears to be negative. The failure of tubewell water and low-lift pump technology that would deliver less than 1-cubic-foot-per-second is a significant anomaly in the whole story of private-sector development. Unlike the Indian Punjab, where the so-called "fractional" water-production technology has often been cited as a key ingredient in the rapid growth that has occurred in the recent past, there seems to have been no response by either the private or the public sector in Pakistan to the demands of small farmers.

Several hypotheses, all subject to further investigation, have been suggested for this result. One, for example, dwells on the economics of the investment in such technology from the farmers' point of view. Gotsch (1978) presents evidence that during the late 1960s and early 1970s, the real costs of low-lift pumps and motors to Pakistan farmers were roughly double those paid by their Indian counterparts.[16]

Although there may be some merit in the argument, substitution of the Indian price data in the Pakistan model does not eliminate mechanical investment in water-producing technology from the small farmer's optimal equipment package. The shadow prices of supplementary water are so high that even at profit rates substantially below those obtained in India, investment in water-producing activities is, at least as far as the model results are concerned, a desirable activity.

Two institutional arguments for the failure of the technology-supply response mechanism to operate have also been advanced. One involves the hypothesis that Pakistani manufacturers simply did not have the know-how to manufacture the smaller, high-speed diesel engines that fractional technology requires. Unlike India, where machine tool capacity was somewhat more highly developed, the Pakistani craftsmen had to rely on the designs that they had been copying for several decades, designs that featured large, cast-iron replicas of 18- to 20-hp diesels they had inherited from the pre-World War II period.

A second suggestion relating to the institutional environment raises much more perplexing questions that have to do with the role of political and social constituencies in determining the types of innovation that are generated. Small farmers form a less significant proportion of the total rural population in Pakistan than they do in India. From a market point of view, they thus represent a smaller opportunity, and it could be argued that Pakistan's small-scale manufacturing sector, even if it had had the requisite skills for producing smaller machines, would have preferred to concentrate its efforts on selling to farmers whose holdings were sufficiently large to utilize the capacity of the bigger machines. On the public-sector side, the potential political power of the small farmer class was overshadowed by that of the "kulak" class of 25 acres and above, with the result that no government efforts were expended on behalf of the small farmers.[17]

The diffusion of water-producing technology was the first in a series of mechanical innovations that have characterized Punjab agriculture over the past·two decades. In addition to the insights it offers into the way in which specific resource scarcities can influence the generation of new mechanical technology, it also emphasizes the dynamic nature of the process. As Table 3 clearly shows, relaxation of water constraints in the benefits matrix has predictably produced a new set of scarce resources. Land has become a binding constraint, which suggests that research devoted to changing the planting and harvesting dates in such a way that double cropping can be carried out will have a high payoff. Similarly, the model's labour constraints have become binding even on small farms in the crucial threshing period, suggesting that specific means should be sought to reduce either the labour needed for harvested crops or the labour needed for crops that compete with one another for labour because of planting requirements. In the following section, a more elaborate model is tested to examine behaviour of the benefits matrix in the presence of a number of technological interactions.

TRACTORS AND THRESHERS IN PUNJAB AGRICULTURE

Step I: Benefits Matrix with Tubewell Technology

THE ACTUAL BENEFIT MATRIX ELEMENTS at the end of Round 1 (Table 3) can be thought of as the elements for the *ex ante* benefits matrix that determines the demand for technology in Round 2. Previous comments have pointed to the significant changes in the resource shadow prices on farms that were able to gain access to supplementary water supplies. Additional land now has a very high scarcity value, an indication that there is active competition among crops for this resource. Anything that would relax this constraint, be it more land or adjustments in the cropping pattern that permitted double cropping, would have a high payoff. Similarly, anything that would release the labour and power constraints in binding months would add substantially to the value of the objective function (i.e. to profits).

The simple fact that a resource constraint is binding does not, or course, indicate the precise nature of the technological package that might be introduced to break the bottleneck. Indeed, that is the issue around which much of the debate concerning "appropriate" technology centres. For example, in the Pakistan case, the binding land constraints shown in Table 3 could conceivably be relaxed by introducing: (1) a wheat variety that matured earlier or cotton and rice varieties that could be planted later; (2) threshers that facilitated the harvesting of wheat; (3) tractors that speeded up seedbed preparation for the summer crops; or (4) pest control practices that permitted later planting of summer crops.

The optimal packages of mechanical technology that would be chosen if farmers were profit maximizers are shown in Table 4.[18] Again, the impact of farm size on the model's optimal package is overwhelmingly apparent. Tractors, threshers, and tubewells, entered as integer variables, are all too "lumpy" to be afforded by small units. With the appearance of hire-service markets, simulated by the model's continuous solution, or with the redesign of the conventional technology, small farmers would still participate to only a limited extent in the widely publicized Green Revolution that has taken place in the area, compared with their medium- and large-sized neighbours.[19]

The model also predicts, however, that there would be significant incentives to the development of hire-service markets in virtually every mechanical technology. The documentation of markets for water was mentioned earlier. One would also expect, from an examination of the activities of the optimal solution, that preparatory tillage by tractors, which permits double cropping, would provide a powerful impetus to the availability of tractor services for this specific task. Similarly, thresher leasing would appear to offer a high return to small farmers since it, too, would contribute to the elimination of labour and animal-power constraints in the spring of the year.

Step II: Induced Bureaucratic and Market Responses

Few South Asian countries, at least in the 1960s, had the capacity to produce tractors in the private sector. Consequently, the response to demands for additional power involved primarily imports from Western countries, usually financed with credits from AID, the World Bank and other foreign aid agencies. These machines were then sold to farmers by government importers at undervalued exchange rates and subsidized credit, a practice that made the tractors significantly cheaper in real terms than they were for farmers in the country of origin.

Several elements appear to have contributed to what was, even at the time, a hotly debated economic policy. For example, as the model results in Table 4 imply, those who were lobbying for the importation of tractors were clearly the larger and hence also the most vocal and influential farmers. Indeed, many members of the civil service were substantial land owners in their own right and undoubtedly saw the availability of tractors as a way of minimizing the labour and management problems of their own operations.

There was also a firm belief, at least on the part of the agricultural establishment, that mechanization was closely associated with modern agriculture. Most of the responsible officials and politicians had travelled extensively in developed countries and the desire to raise their standards to those they had observed must have been strong.[20] Tractors seemed to be the key to this transformation.

Lastly, the Great Tractor Debate did not reach full stride until the middle and late 1960s. It was only with the advent of the Green Revolution and the obvious benefits of resuming land for self-cultivation that the number of inquiries regarding the impact of tractors on labour and tenant displacement were undertaken. Before that, there was not much information on the impact of machines on such social parameters as employment and the distribution of income, at least not in developing countries. As a consequence, in its initial stages, the discussion was often illustrated with information drawn from the experience of developed countries. Bureaucrats intent on continuing the funding of machines under foreign aid loans that appeared to have a low opportunity cost found it relatively easy to discount experience coming from such a disparate source.

Previous comments have noted that in Pakistan there was a significant market response to the demand for water-production devices such as tubewells, low-lift pumps, and motors; and that there was a positive bureaucratic response to the demand for tractors. Curiously, however, there was little or no response through either mechanism to the need for threshers. The model results suggest that medium-sized farmers who could not afford a tractor could have made significant progress in releasing springtime labour and power constraints if they had been able to thresh wheat mechanically. Indeed, should there be doubt about the validity of this calculation, the rapid development of small threshers in East Punjab (India) corroborates the "synthetic" results that benefit-cost ratios for such a machine would, especially among the smaller farmers, be sufficient to cause their rapid diffusion.[21]

Step III: Economic Structure and Adoption

Ultimately, the heated debates about the impact of tractors generated a substantial amount of field investigation. Although the situation still needs further monitoring because of the relatively incomplete character of mechanization,the following major conclusions have emerged:[22]

1. When properly corrected for management bias, the introduction of the tractors has not led to improvements in yields or crop intensities.

2. Those who have purchased tractors have begun to cultivate share cropped land themselves and attempted to increase the size of their operated holding by renting in land. The result has been significant displacement of tenant farmers. With some exceptions, tractors have also reduced the amount of labour used on the mechanized farms.

3. Hire-service markets in tractor services have continued to spread. Unfortunately, there is no recent evidence available from farm management surveys as to the precise dimension of such activities, but casual conversations with farmers in the countryside confirm that many small farmers have decreased the total number of draft animals on their holdings because of the widespread availability of tractors for preparatory tillage and seedbed preparation.

4. Threshers remain an engima. The Government of Pakistan, in cooperation with the International Rice Research Institute, now has a programme aimed at diffusing an axial-flow thresher developed in the Philippines. It is not known how successful these efforts have been to date or the extent to which machines of different designs have come into more widespread use.

The overall impression of the second round of mechanization is that it is following a pattern similar to that followed by developed countries. The difference in the two situations, of course, is that there is little similarity in the global shadow prices of capital and labour. Obviously, other elements having to do with the institutional and political environment in which change is taking place are a significant impact on the induced-innovation process.

The dotted line on the right side of the transnational systems diagram (Figure 2) is meant to suggest that this latter fact is now much more clearly recognized by certain segments of the international community. Indeed, the debates *within* several foreign aid agencies have been rather acrimonious as staff economists argued with project specialists about the merits of tractor loans, infrastructure investments and the like. Although there is considerable evidence that some significant changes in project selection have resulted, the overall impact of an increased awareness of the potentially adverse distributive consequences of mechanization remains to be seen.

DIRECTIONS FOR FURTHER RESEARCH: A COMPARATIVE APPROACH

JUXTAPOSITION of the broad systems framework of Section 2 with the detailed materials available from the Pakistan experience is both heartening and discouraging. The positive finding is that the task-specific programming investigation of the farming system provided important insights into the types of innovations that would be generated if farmers had their way. Of particular interest in terms of giving the benefits matrix content were the difference—and similarities—that existed between farm size groups regarding what each would regard as "appropriate" technology.

The discouraging aspect of reviewing the Pakistan experience was the shallowness of the material on the technology response side. Although an attempt was made to "complete the system", little of the material used had the kind of solid empirical base that was available on the demand side. The result was to suggest clearly the desirability of according a high priority to work on linkages and the inducement mechanism.

Individual case studies that related expenditures on the development and diffusion of new technology to particular commodities, regions and social groups, would expand current understanding of short-run inducement mechanisms. These insights would be further enhanced, however, if they could be carried out within a comparative framework. While the systems approach is helpful in identifying interactions and organizing an overall perspective on a problem, its very comprehensiveness causes difficulties of causal interpretation. Although it is obviously impossible to achieve the "control" of the natural scientist, *appropriately* chosen comparative studies can assist in establishing causality by employing the plausible assumption that certain parameters can be held "constant" between systems.

In the case of agricultural technology, the most critical "control" is the agro-climatic environment that provides the context for the farming system being investigated. There is not much point, for example, in comparing the benefit matrices for humid, tropical agriculture and arid, irrigated agriculture to develop clues about the differences in institutional response. The determinants of scarce resources in these two systems are so different that little can be added to the individual analyses of system behaviour. On the other hand, where agricultural environments do exhibit considerable similarity in precipitation, soils, crops grown, technical coefficients, etc., a comparison across environments of the characteristics of the technology and its institutional environment yields important insights into the way these latter parts of the system influence the choice and diffusion of new technology.

Two areas that lend themselves to comparative analyses have already been mentioned, i.e. the Indian and Pakistan Punjabs. To these might be added a third, namely, the Nile Delta of Egypt. The latter is also wholly irrigated, dependent upon a combination of animals and tractors for power, is a producer of rice, cotton, wheat and maize, and has a high cropping intensity. The experience of the country, however, has been significantly different in terms of its pricing policies, attitudes towards agricultural mechanization and, most importantly, the role of government in all phases of the economy. The following paragraphs can only touch on some of the issues in which comparative case studies could be undertaken. There is sufficient casual evidence to suggest, however, that: (1) the condition of similar agro-climatic environments on the demand side is fulfilled; and (2) that the responses to the same "signals" for improved technology have produced very different responses on the supply side. Succeeding sections both deliniate questions and hazard answers about what a more detailed investigation would show.

INNOVATION IN THE PRIVATE SECTOR

ANYONE WHO HAS VISITED the Indian Punjab comes away impressed with the transformation that has occurred there in the past two decades. Much has been written, of course, regarding the impact of high yielding varieties, irrigation, and mechanization. Most knowledgeable visitors, however, are even more impressed by the industrial transformation that has taken place in the small towns and villages. Each seems to have an astonishing variety of machine shops and other small-scale industrial establishments that are producing a wide range of both producer and consumer goods. Of particular interest are the shops producing agricultural machines, both for use in the Punjab and for export to other parts of India.

The previous description of developments in Pakistan has indicated that there, too, one finds a thriving small-scale manufacturing sector. Although questions have been raised about its capacity to innovate, particularly in water producing down-sizing and post harvest machines, the private sector has nevertheless made an obvious contribution to developments in the agricultural sector.

In Egypt, however, the private sector is virtually non-existent. Only in the past several years have small manufacturers begun to respond to the needs of farmers and then largely under the stimulus of outside organizations. (One of the most innovative has been the programme initiated by the Ford Foundation and the Catholic Relief Service to "create" the demand for a small, axial flow thresher of the type developed by International Rice Research Institute (IRRI) in the Philippines.)

One could conjecture about the sources of the differences in the role of private sector enterprises in the three areas. However, what is missing in terms of a solid basis for comparative work are studies on the economics of light industrial establishments. At a minimum, the following kinds of data are needed:

1. Input/output relationships for producing pumps, threshers, tractor implements, motors, etc.
2. The costs of raw materials used in the production of various machines.
3. Wages paid to labour.

4. Constraints of an institutional nature, e.g. government restrictions on prices or on the importation of locally unobtainable materials and parts.
5. Subsidies on completed implements.
6. Miscellaneous costs, especially expenditures associated with advertising and promotion.

Much of this information could be collected by questionnaire in much the same way that farm management data are obtained. On the basis of the results, budgets for the profitability of manufacturing and selling various types of machines could easily be generated and issues of skill, information and economics could be disentangled.

The motivation for this type of detailed questioning is, of course, its ability to illuminate the larger differences in the private sector performance observed in the three countries being compared. What type of hypotheses might one have in mind? First, on the basis of previous work on the Indian and Pakistan Punjabs, there is reason to suspect that substantial differences have existed in the area of raw material procurement. This is not only a matter of price policy with respect to pig iron and scrap, but with the way in which the entire industrial licensing system affected small-scale industry in the two regions.

Government policies towards private enterprise in general are also likely to be important in explaining the failure of a small-scale industrial sector to emerge in the Egyptian case. While India was experiencing a kind of chaotic innovation process based on private profitability, during the Nasser period, the mere suggestion of private profitability was sufficient to produce nationalization in Egypt.

Pakistan probably falls somewhere in between the two previous examples. Although there were no specific hindrances placed in the way of small-scale industries, the industrial policy generally and agricultural mechanization subsidies in particular that were in effect during the 1970s were not sufficient to produce the proliferation of new firms offering new products that characterized the Indian Punjab.

INNOVATIONS GENERATED IN THE PUBLIC SECTOR

THERE ARE TWO TYPES of related issues that would seem to be worthy of a comparative investigation. First, there is the question of how the present system of public sector research functions and the incentives that are operating on research scientists and research administrators. Second, there is the further question of the forces that have produced the situation in successive "rounds" of systems behaviour.

Immediate Research Incentives

As Ruttan notes in his most recent paper on induced innovation processes:

> The response of research scientists and administrators represents the critical link in the inducement mechanism. The model does not imply that it is necessary for individual scientists in public institutions to consciously respond to market prices, or directly to farmer's demands for research results in the selection of research objectives. . . . It is only necessary that there exists an effective mechanism to reward the scientists or administrators, materially or by prestige, for their contributions to the problems that are of social or economic significance.

Data on immediate incentives to researchers in developing countries are not terribly hard to obtain. They are admittedly not the sort of thing that one can find in internationally published sources, but they are generally available through personal interviews. A recent AID Science and Technology mission to Egypt, for example, found readily accessible material on disaggregated research budgets, salaries, equipment availability and, in academic institutions, teaching responsibilities. The result was a picture of incentives operating on individuals that left no doubt as to some of the reasons for the type of research that was being done and the limited productivity of what is otherwise a large, rather well-trained cadre of agricultural scientists. It had nothing to do with the complaint sometimes heard about "inappropriate Western training" and much to do with government policies towards research, education and employment in general.

The type of information indicated above would also be available in Pakistan and India. It is, for the most part, a matter of public record. However, grubbing will be necessary. Local publication does not imply widespread accessibility.

The "Political Economy" of Research

Ruttan's work (1980) again puts the question of how public sector resources are allocated in perspective:

> The relative power of different economic and social groups over the politico-bureaucratic structure is the primary determinant in getting their specific demand eventually translated into a supply of new knowledge or new technology. In the case of technology, pressure on the politico-bureaucratic structure results in a specific allocation of funds and of human capital research institutions and within these, to particular lines of research.

What kinds of data would provide insights into the nature of the linkages between various social groups and the agricultural research establishment? The following materials, also readily available in most research organizations at both the individual station and national levels, would be relatively easy to equate with influences from the demand side.

1. Expenditures by commodity: Estimates could be used to construct a series of ratios with the value-added by that commodity to the country's agriculture. Such ratios would, where they diverge from some sort of rough "parity", be extremely useful in identifying, by implication, a particular set of growers or consumers as being the primary beneficiaries of research.

2. Expenditures by geographical region: Data organized by regional research facility would be extremely useful in identifying the importance of space (and who owns it) on research allocations.

3. Expenditures by discipline: Despite their increasing orientation toward a project concept of research, most research organizations in the three countries proposed for comparison continue to display strong disciplinary compartments. Figures on the amount spent, for, say, agricultural engineering versus plant breeding are reasonably easy to obtain and would offer additional insights into the characteristics of the economic and social groups to whom the research establishment as a whole was responding.

Identifying the inducement mechanisms of a particular system is obviously only a first step in a long and arduous process of improving the efficiency with which research resources are allocated. However, in our judgement, developing these estimates within a framework of similar situations is likely to have a greater policy impact than additional exhortations from scientists and administrators in advanced countries or international research stations.

Notes

[1]Compare materials reviewed in Hans P. Binswanger's "Induced Technical Change: Evolution of Thought", in Binswanger *et al.*(1978), with Hightower (1973).

[2]For somewhat different systems approaches to the issues involved, see de Janvry (1977), Gotsch (1974), Gemmill and Eicher (1973), and Ruttan (1978).

[3]These are of course the economist's "shadow prices" and the programmer's "dual variables".

[4]For a model in which both costs and benefits of political activity are examined within the context of a general equilibrium economic model, see Andersen (1977). The issues arising at this point in the system are also similar to those addressed by Ruttan under the heading of "induced institutional innovation". (Cf. Binswanger *et al.*1978).

[5]Criticisms of this sort have been frequently directed at the system of International Agricultural Research Centres that has developed during the past two decades. In response, substantial changes have taken place in both the scientific approach and outreach relationships with national research programmes.

[6]It is interesting to contrast the effectiveness of agricultural research organizations in certain regions of India where the regional influence has been strongest, e.g. in Punjab and Haryana, the McCarthy findings have, in the case of agriculture, been strongly supported.

[7]See, for example, Villamil (1977) for some of the recent literature on "dependency" theories of Latin American development.

[8]In this category, the multi-national corporation has of course been the target of the greatest criticism.

[9]For a number of studies employing the programming methodology, see Gotsch, *et al.* (1975).

[10]A Persian wheel, the traditional means of obtaining additional water in the Punjab, uses the tractive power of animals walking in a circle to drive an endless chain of buckets that lift water from a shallow percolation well dug some 20 to 50 feet below the surface.

[11]The example of groundwater development in Pakistan illustrates a case in which considerable tension existed between public and private interests. The difficulty arose when the government sought to prevent individual wells from being drilled in areas where it was thought public projects would ultimately be located. Subsequently, this policy was abandoned and public projects (the so-called "SCARPs") now contain both private and public wells. As Hightower (1973) has argued, this type of conflict between public organizations in agriculture (e.g. the land-grant colleges and government research institutes) and dominant private interests must be considered the exception and not the norm.

[12]This concept was a major contribution of the White House Panel, *Report on Land and Water Development in the Indus Plain* (1965), the so-called "Revelle Report".

[13]The modified model incorporating tubewells and high-yielding varieties is described in Gotsch *et al.* (1975), pp. 27-46.

[14]Indeed, as Bashir Ahmed's cross-sectional survey shows, small sharecroppers were often displaced entirely from the land because their landlord wanted to retain the entire increase in productivity from the tubewell for himself (Ahmed, 1972).

[15]These results are verified by Mohammed Naseem's (1971) cross-sectional survey of farms in Sahiwal District shortly after the Green Revolution got under way. Not only did he find that farmers who purchased water used less than farmers of a similar size who owned a tubewell, but

that the difference was consistent with that predicted when water prices were substituted for assumed tubewell capacity in his programming model.

[16]The benefit-cost ratios that support this argument were affected by both higher output returns and lower (subsidized) pump costs in India.

[17]Several authors have advanced this point of view, including Gotsch (1978) and Sen and Amjad (1977). The most definitive work along these lines is contained in Hamid (1980). In all cases, however, the link between the social and political pressures alleged to have been exerted on the bureaucracy are only suggestive, and little empirical evidence has been brought to bear on the issue.

[18]Only mechanical solutions to the relative factor scarcities are shown. However, the model also contains technology variables for such divisible inputs as HYV and pest control. Because these types of technology tend to dominate traditional practices, they have not received additional attention. The mixed-integer model on which Table 4 is based is described in detail in Gotsch *et al.* (1975).

[19]As noted earlier, the evidence strongly suggests that improved seeds and fertilizers are being used extensively and in roughly the same measure by large and small farmers alike. What is at issue in this paper is access to mechanical technology. For a comprehensive review of the small-farmer experience in Asia, see Singh (1978).

[20]Many of these same individuals, of course, also received their technical training in agriculture in the United States during the 1950s, when USAID financed a large number of Ph.D.s for young government officials from developing countries.

[21]The thresher case is such an anomaly in the Pakistan scene that it would appear to be worth further investigation. Hypotheses similar to those invoked to explain the absence of a market response to fractional water-producing technology might have some validity. However, in this case, mechanical threshing would appear to be advantageous to medium and small farmers alike; consequently, the limited-size-of-the-market argument does not seem persuasive. A more likely hypothesis is that, without a crucial R&D breakthrough—a breakthrough that occurred in East Punjab when a simple fodder chopper was fitted with a hood and a blower—the knowledge of how to construct such machines simply did not exist. Without prototypes, the relatively unsophisticated machine-tool industry had nothing to copy and did not have the information and the skills to respond from its own resources.

[22]The major conclusions regarding tractors appear to hold in both the Indian and Pakistan Punjabs. For a detailed review of the literature on tractors in Brazil for comparison, see Binswanger *et al.* (1978), Chapter 10.

References

Ahmed, Bashir. (1975): "Tractor Mechanization in the Punjab", in *Food Research Institute Studies,* Vol. XIV, No. 1.

Anderson, K. (1977): "Distributional Aspects of Trade Protection in Australia with Emphasis on the Rural Sector", Ph.D. Dissertation, Food Research Institute, Stanford University.

Binswanger, H. *et al.* (1978): *Induced Innovation.* Baltimore, Johns Hopkins University Press.

Day, R.H. and Inderjit Singh (1977): *Economic Development as an Adaptive Process, The Green Revolution in the Indian Punjab.* Cambridge, Cambridge University Press.

de Janvry, A. (T.M. Arndt *et al.* eds.) (1977): "Inducement of Technological and Institutional Innovations: An Interpretive Framework", in *Resource Allocation and Productivity in National and International Agricultural Research.* Minneapolis: University of Minnesota Press.

de Janvry, A. (1978): "Social Structure and Biased Technical Change in Argentine Agriculture", in *Induced Innovations,* edited by H.V. Binswanger and V. W. Ruttan. Baltimore, Johns Hopkins University Press.

Dunkerley, M.P. (1979): "Choice of Technologies: The Influence of Multilateral Agencies", in *Appropriate Technologies for Third World Development,* edited by A. Robinson. London, The Macmillan Press.

Gemmil, G. and C. Eicher (1973): "A Framework for Research on the Economics of Farm Mechanization in Developing Countries", Rural Employment Paper No. 6, Department of Agricultural Economics, Michigan State University, East Lansing, Michigan.

Gotsch, Carl H. (1974): "Economics, Institutions and Employment Generation in Rural Areas", in *Employment in Developing Nations,* edited by E.O. Edwards. New York, Columbia University Press.

Gotsch, Carl H. (1975): "Linear Programming and Agricultural Policy: Microstudies of the Pakistani Punjab", *Food Research Institute Studies,* 14, No. 1.

Herrera, A. (1973): "Social Determinants of Science Policy in Latin America: Explicit Science Policy and Implicit Science Policy", *Journal of Developing Studies.*

Hamid, N. (1980): Forthcoming Ph.D. Dissertation, Department of Economics, Stanford University.

Hightower, J. (1973): *Hard Tomatoes, Hard Times.* Cambridge, Schenkman Publishing Company.

McCarthy, F. (1972): "Third Cultural Networks of Philippine Social Scientists", Ph.D. Dissertation, University of Michigan, Ann Arbor, Michigan.

Naseem, M. (1971): "Small Farmers and the Agricultural Transformation of West Pakistan", Ph.D. Dissertation, University of California, Davis, California.

Ranis, G. (1979): "Technology Choice and Employment in Developing Countries: A Synthesis of Economic Growth Centre Research", Report to AID, Yale University, New Haven.

Ranis, G. (1979): "Appropriate Technology: Obstacles and Opportunities", in *Technology and Economic Development,* edited by S.M. Rosenblatt. Boulder, Westview Press.

Rogers, E. M. and Shoemaker, F.F. (1971): *Communication of Innovations, A Cross-Cultural Approach.* New York, The Free Press.

Ruttan, V.W. (1975): "Technical and Institutional Transfer in Agricultural Development", *Research Policy, 4.*

Ruttan, V. W. (1978): "Induced Institutional Innovation", in *Induced Innovation,* edited by H. Binswanger *et al.* Baltimore, Johns Hopkins University Press.

Ruttan, V. W. (1980): "Institutional Technology: Issues, Concepts and Analysis", Working Papers, World Employment Programme Research, ILO, Geneva.

Sen, A. and Amjad, R. (1977): "Limitations of a Technological Interpretation of Agricultural Performance: A Comparison of East Punjab (India) and West Punjab (Pakistan) 1947 - 1972", *South Asia Papers,* Vol. 1, No. 11-12, University of Punjab, Lahore.

Singh, I.J. (1978): "Small Farmers and the Landless in South Asia", Background Paper 4, *World Development Report.* The World Bank, Washington, D.C.

Stewart, F. (1977): *Technology and Underdevelopment.* London, Macmillan.

Sunkel, O. (1971): "Underdevelopment: The Transfer of Science and the Latin American University", *Human Relations, 1.*

Villamil, J. J., ed. (1977): *Transnational Capitalism and National Development.* Atlantic Highlands, New Jersey, Humanities Press.

CHAPTER 3

Determinants and Effects of Technological Choice

Jan Svejnar
Assistant Professor of Economics
Cornell University

and

Erik Thorbecke
M.E. Babcock Professor of Economics and Food Economics
Cornell University

INTRODUCTION

THE FACT THAT the choice of technology affects crucially a country's economic, social and political development is now generally accepted. At the same time, in spite of numerous studies in the area, there appear to be at least three questions which are yet unresolved: (1) the formulation of a framework within which the major institutional and other factors that determine the state of technology in a given country setting can be analyzed; (2) the design and application of a decision-making model explaining the technological choice process through the interaction of different agents; and (3) the formulation of a methodology to incorporate the technological dimension into a macro-economic framework to evaluate analytically the effects of choice on relevant policy variables. An additional issue which, in a sense, relates to the preceding ones, is to define more rigorously the concepts of "technology" and "techniques" from different disciplinary vantage points, i.e. economic, engineering and socio-political.

In order to tackle successfully the three main issues identified above, a research approach needs to combine institutional knowledge with appropriate technical information. In addition, conceptual models incorporating the institutional and technical data have to be developed so as to analyze the factors affecting the choice of technology and the impact of any particular choice on policy variables.

63

The research which is currently being carried out by the Cornell team proceeds along these lines. Over the past several years Cornell's programme on Science and Technology has engaged in a considerable data collection effort with respect to various institutional settings. During the past year this ongoing activity has focussed on the following countries: The Republic of Korea, Malaysia, Colombia, Mexico and Nigeria. At the same time intensive model-building efforts have been initiated with the aim of producing a conceptual framework to help explain the decision-making process regarding technological choice and how the latter, in turn, affects major development objectives at the macroeconomic level.

The primary aim of the country studies undertaken by the Cornell team was to try to shed some light on the first question above by examining carefully the institutions relevant for an empirical analysis of the choice of technology. Since each country pursues a specific form of development, the relevant economic structures, institutions and decision-makers differ from one country to another. In addition, knowledge of these particular institutional settings is, of course, required to help answer the relevant analytical models such as those described later in this chapter.

The present chapter focusses on the second and third issues mentioned above.[1] We survey briefly and selectively, by way of background, the main existing analytical literature on choice of technology. It contains a partial equilibrium decision-making model which can be used to analyze and predict technical choices in situations where more than one agent influences the outcome. This section also elaborates on several intended applications of the decision-making model in the next phase of our research. The final section presents a conceptual framework based on the Social Accounting Matrix which incorporates technology and can be used to explore the macroeconomic effects of alternative technologies. This framework is applied to the case of South Korea for illustrative purposes.

Existing Literature

The purpose of this section is not to survey exhaustively the literature on choice of technology. Rather our aim is to discuss briefly the main areas in which analytical work has been done to date and indicate where our work, described in the following sections, fits in.

The first and surprising finding is that virtually no analytical work has been done at the macroeconomic level in trying to incorporate technology within a comprehensive and empirical intersectoral framework to estimate the effects of alternative technologies on such development objectives as output, income distribution and employment. A model based on the Social Accounting Matrix is presented in this chapter and appears to be one of the first attempts at exploring the macroeconomic effects of technological choice within an empirically-based general equilibrium framework.

The microeconomic literature has focussed primarily on the observed "paradoxes" of inappropriate technology choices. In view of the different factor endowments and factor prices in the developed and less developed countries (DCs and LDCs), the standard behavioural models of profit maximization or constrained cost minimization predicts that more labour-intensive techniques will be used in the LDCs than in the DCs. The technologies which are observed in use, however, seem to be inappropriate in terms of the standard model and three basic paradoxes are observed in the empirical literature:

> (a) A capital-intensive technology is chosen which is distinctively more costly than a relatively less capital-intensive one and differentials in product quality cannot justify such a decision (See for instance Y. W. Rhee and L. Westphal, 1977);

> (b) Several techniques, ranging from traditional and relatively labour-intensive to very modern and relatively capital-intensive are observed in use within the same industry and within the same region (this finding is reported among others by L. Wells, 1975);

> (c) There is a differential in the rate at which improved technologies are adopted by producers in the same sector and region, which results in the phenomenon observed in (b) (C. Bell's 1972 study of the acquisition of agricultural technology is a case in point).

The observed phenomena are at best counter-intuitive from the standpoint of a standard microeconomic model. A policy concern over the appropriateness of the technological choice therefore demands a better analysis of the decision-making process and factors, which lead to the observed technological outcome.

Unfortunately, little analytical work has been done in this realm to date. The literature can be basically divided into three broad areas comprising three different approaches:[2]

> (i) Examination of the possibilities of substitution of factors within production. Is there a set of feasible techniques to be chosen from or is technology inflexible?

> (ii) Suggestions that the standard model is still a good description of behaviour, but that the objective function is constrained by additional factors which can cause the cost-minimization outcome to differ from the actual one. It is argued that these factors vary across agents and thus explain the wide range of techniques observed within sectors and industries;

> (iii) Proposals for a totally new or modified decision-making model, i.e. one that differs from the constrained cost-minimization (profit maximization) model for producer behaviour.

Category (i) is characterized mainly by studies of the elasticity of input substitution in production. In most cases this means capital-labour substitution along a postulated isoquant[3]. Although some of these studies suffer from methodological problems, there is a general consensus that in most cases a choice among techniques does exist. It is also agreed that there is more flexibility in peripheral aspects of production than in core aspects. The main point to be derived from this literature is that an "inappropriate" choice, in general, cannot be explained by the existence of only one technique of production. Within category (ii) we find studies examining the various phenomena which are thought to explain the seemingly inappropriate technology choices. Risk and different degrees of risk aversion among small and large producers are regarded by some researchers (e.g. see C. Bell, 1972) as a cause of variation in the capital-intensity. This approach naturally incorporates the idea that technology is not "scale neutral". Credit constraints and market power are both cited as favouring the larger firms and inducing them to select more capital-intensive modes of production (see L. Wells, 1975). Multinational corporations (MNCs) are often regarded as rigid with respect to the choice of technology and automatically adopting the capital-intensive technology from DCs[4]. Finally, government regulations with respect to trade, exchange rates, taxes, interest rates, and labour laws are claimed to provide sufficient price distortions to explain the observed choice of technology (e.g. L. Westphal and Y. W. Rhee, 1977).

To our knowledge category (iii) has so far been represented by two alternative models of the decision-making process. Gershon Feder (1980) uses an expected utility maximization model to examine the issues of farm size and adoption of new technology. Utility is assumed to be a strictly concave function of income only, the concavity assumption reflecting risk aversion. Moreover, the situation is characterized by uncertainty with respect to the quantity of output. The aim is to develop a model of decision-making in the agricultural sector which could be used to examine the problem of the optional choice among risk activities under expected utility maximization. Farmers choose between a modern technology and a traditional one, the former characterized by relatively more risk, more inputs (e.g. fertilizer), and the potential for higher yields. The analysis tries to clarify rigorously several phenomena which are thought to impinge on the producer's decision to adopt a technology and the rate at which it is adopted: risk, risk aversion, farm size, and credit constraints. The interesting policy result that Feder obtains from the reduced form of his model is that the optimal level of fertilizer per acre (which together with other variables represents the optimal adoption and use of a more advanced technique) is independent of: (a) the degree of risk aversion; (b) the farm size; and (c) the degree of uncertainty about the output level (i.e. the variation in the random variable determining the variability of output).

R. Day and I. Singh (1977) developed a recursive model of "an adaptive, multi-goal theory of decision-making that incorporates realistic behavioural rules and suboptimization with feedback". The model is developed for the agricultural sector and farmers are characterized as using cautious, optimizing techniques. "Activities that are feasible from technological and financial points of view are further circumscribed by adaptive constraints that accommodate the cumulative effects of experience and observations of neighbours' experience." In addition farmers have explicit multiple goals which do not have equal priority. The model postulates four goals as those to be "attained" in order of priority in a lexicographic ordering: a subsistence level, a level of cash consumptions, safety (incorporating risk), and profit maximization. The first three are satiable. The farmer chooses from among technologies by satisfying these first three criteria in order, and then by choosing which technology will maximize profit. Furthermore, the farmer's decisions are continually modified through a feedback mechanism.

The area of decision-making models clearly constitutes an enormous potential for a better understanding of the choice of technology. The recognition that an agent may have a multiplicity of goals is especially important. Research efforts in this direction greatly increase the realism and hence the policy relevance of the analytical work in this field. The decision-making model presented in the next section allows for multiple goals. In addition, it is designed to cover the cases when several decision-makers co-determine the observed outcome (choice). Since in real-world situations a single agent rarely selects a technology without being influenced by others, we feel that our approach provides a needed addition to the existing stock of analytical tools in the area of technology choice.

Decision-making Model

In this section we develop a game-theoretic approach for analyzing the outcomes of "group decisions" with respect to the choice of techniques. Most outcomes of choice in this area result not from the decision of a single person (e.g. a farmer, extension agent, entrepreneur) or an institution, such as government, union, board of directors, bank, or the headquarters of a multinational corporation (MNC), but rather from a complex process of bargaining in which usually several agents strategically interact among themselves until they finally "agree" on an outcome that is acceptable[5] to all of them. In a game-theoretic framework we are therefore observing the outcome of a "cooperative game" in which the relevant agents can and do exchange information (farmer with the extension agent, MNC with the local government, etc.) and search for a jointly acceptable solution.

We will presently describe the general model. In the following sub-section we shall then indicate how the model can be applied in specific institutional settings.

We consider an environment characterized by a vector G of ι goals:

$$(3.1) \qquad G = G_1, G_2, \ldots, G_\iota),$$

and a vector N of n agents (co-decision-makers), $N = (N_1, \ldots, N_n)$. When $\iota = n = 1$ the situation reduces to the traditional framework of one agent with one goal (e.g. profit maximization). In general, a particular agent may be interested in the achievement of one or several of the goals in G. In the latter case each agent will presumably use his/her bargaining power (degree of influence) so that more preferred goals receive more weight than those which are of a lower priority. To characterize this process more clearly let us introduce a vector Γ of ι weights $\Gamma = (\gamma_1, \gamma_2, \ldots, \gamma_\iota)$, corresponding to the vector of goals G. Given the environment (endowments, prices, etc.), vector Γ will be shown to determine the degree to which the various goals in G are fulfilled ex post-facto. If each agent is interested in only one goal and no two agents are intended in the same goal, then $\iota = n$ and γ_i is the weight' attached to goal G_i and reflects the total bargaining power of agent i. If several agents have a preference for any given goal, the weight assigned to that goal should reflect the preferences and relative bargaining power of all who advocate that goal. To make these concepts simpler and operational it is convenient to normalize Γ so that

$$(3.2) \qquad \sum_{i=1}^{\iota} \gamma_i = 1 \text{ and } 0 < \gamma_i < 1 \text{ for all i.}[6]$$

Each agent divides his total bargaining power (influence) in support of various goals. The weight γ_i of G_i is then equal to the sum of the weights provided by all the individual agents for goal G_i. For example, suppose that G_i is advocated by agents N_1 and N_2 only. If N_1 provides a weight of 0.1 and N_2 of 0.05 behind G_i, then $\gamma_i = 0.15$.

Individual decision-makers usually have a good idea about the goals whose fulfilment they advocate and about the relative emphasis they would individually place on their goals. Some decision-makers may be even willing to convey this information about their goals and relative weights. Others, however, may be unwilling or simply unable to describe their weighted preferences. Moreover, even if we obtain the revealed information, unless we know the relative bargaining power (degree of influence) of each of the agents, we are still unable to aggregate this information to obtain Γ. At the same time the knowledge of Γ is clearly important because it tells us to what extent the various goals (growth, profit maximization, employment generation, etc.) are pursued and the technologies (techniques) appropriate for the fulfilment of these goals therefore selected. As a result, given the technological options and the various goals, a crucial research task in this area becomes the estimation of the vector Γ. From the policy point of view, apart from variations in the standard environmental variables (endowments, prices, etc.), it is the changes in Γ that lead to the corresponding changes in the fulfilment of various goals through the adoption of alternative technologies (techniques).

In order to know something meaningful about the determinants of the technological choice, we thus need a theoretical model, which would clearly relate Γ to G and, given other environmental factors, provide direct testable prediction. Within the outlined cooperative game approach it is possible to derive a functional form which determines uniquely the outcome of this choice. The result is derived on the basis of the following four axioms.

A 1 In any given situation the decision-makers make a (unique) choice which leads to an observed outcome;

A 2 The outcome is efficient by the Pareto criterion;

A 3 The degree of fulfilment of $G = (G_1, \ldots, G_l)$ is measurable in terms of money or some other measure (e.g. tons of rice).

A 4 The outcome, in terms of fulfilment of G, is proportional to Γ.

A 1 simply states that in a given situation the decision-makers will select technologies and techniques so that we observe a given fulfilment of the goals G. This axiom therefore guarantees that even theoretically we deal with a unique outcome (which corresponds to the observed reality) and hence go beyond the indeterminacy of the classical economic approach to strategic interaction. Note that all the other approaches to technological choice achieve a theoretically unique outcome only by assuming away the problem of strategic interaction (i.e. a single agent makes the decision). A 2 is an efficiency postulate and states that the outcome is such that the fulfilment of any single goal cannot be increased without decreasing the fulfilment of some other goal. Other efficiency postulates such as individual rationality could be used instead (see A. Roth, 1977). A 3 permits a common denominator in measuring the degree of fulfilment of each goal. A 4 states that if $\Gamma_1, \ldots, \Gamma_l$ are the weights that $G_1, \ldots, G_l$ are given, respectively, then the goals ought to be fulfilled in proportion to their respective weights.

On the basis of these four axioms we can prove the following proposition (for proof see Appendix to this Section):

Proposition: If conditions A 1 to A 4 hold, the outcome of the decision-making process can be characterized by the maximization of the objective function V:

$$(3.3) \qquad V = \prod_{i=1}^{l} G_i^{\gamma_i}$$

Formally the decision-making process can therefore be described as a maximization of the weighted product of the goals, where the exponential weights, γ's, are the relative priorities that the decision-makers jointly give to the individual goals. In the following sections we show a few proposed applications of the model.

Applications of the Decision-making Model

The treatment of the decision-making model has so far been at a fairly high level of abstraction. In order to make the model operational it is necessary to define the goals and embed them into a specific environmental setting. It is important to realize that the model is a decision-making structure which can be applied in situations that vary in terms of the number of decision-makers and goals, the range of technological options available, factor endowments, and the economic and other parameters of the system (e.g. prices, tariffs, quotas and other constraints). The empirical research task therefore consists of identifying all these features in a given setting (e.g. an industry) and then applying the decision-making model in order to estimate the weights that the various goals in fact receive in the decision-making process.

For expositional purposes it is convenient to divide the analytical work into two stages. In the first stage, information on available technologies (techniques), factor endowments and the various quantity constraints gives us the technological (input-output) options space. The vast majority of the existing literature satisfies this stage by postulating a well-behaved production function and by collecting data on inputs and outputs. In the second stage one combines the technological options with economic data on prices and other constraints in order to draw conclusions about: (a) the options that a decision-maker has in terms of pursuing an economic goal such as profit maximization; and (b) which technique(s) (technologies) maximize the decision-maker's objective the most. These two steps are known to us from the standard literature reviewed earlier. The decision-making model advances the state of the art in that the technological options and the price information can now be combined with a multiplicity of decision-makers and goals, and that analytically there is a one-to-one correspondence between the technological option selected (observed) and the relative fulfilment of the various goals. Hence, if we can specify the technological options, the other existing constraints, and the objectives (goals) of the relevant decision-makers, then as we observe (collect data on) technological choices we can make inferences about the actual weights that are given to the various goals. Notice that methodologically the only additional information needed for operationalizing this approach is the information on the various relevant goals. All the other information (technological options, constraints, etc.) are needed for the traditional approaches as well. This is, of course, not to say that the existing literature has done justice to these other aspects of analytical inquiry. In fact, considerable additional work in these areas is needed. The point which is being made here is that the decision-making model can provide considerably more insight and policy significance into the issue of technological choice without requiring the collection of a vast additional body of information.

As an illustration, a hypothetical textbook example is now provided to indicate how the model operates in a neoclassical framework once the goals are specified. Following this example is a brief description of how the model can be used to analyze empirically the choice of technology in several specific areas.

Consider a simple hypothetical case when the firm attempts to maximize profit, $\P$, and the government or some other institution (e.g. a trade union) promotes the maximization of employment, $\bar{L}$, as a goal. Both agents have the power to discontinue or impede the operation of the firm and the situation is hence one of strategic interaction. Will the selected technology and/or technique be different than under simple profit maximization? Let the neoclassical production function,

$$Q = Q(L,K), \quad Q_L, Q_K > O, \quad Q_{LL} \leq O, \quad K = \text{capital input,}$$

represent the available technology or an envelope of technologies. The technology choice then reduces to the choice of a technique. Let the profit function be given by $\Pi = PQ - WL - rK - \Pi_o$, where P = output price, W = per unit labour cost, r = per unit cost of capital, and Π_o = the minimum profit level the firm is willing to accept rather than close down. Similarly, let L be measured as an increment over $\bar{L}$, some basic level of L that the other agent (e.g. government or trade union) requires as a minimum for not closing or impeding the operation of the firm. The amount $\bar{L}$ can be any non-negative number (including zero). The objective function V is then given by:

$$(3.4) \quad V = \Pi^{\gamma} L^{(1-\gamma)} = (PQ(L,K) - WL - rK - \Pi_o)^{\gamma} (L)^{(1-\gamma)},$$

where γ is the bargaining power of the firm and $(1-\gamma)$ is the power (influence) of the other party. Note that when $\gamma = 1$ $(1-\gamma = 0)$ the problem reduces to that of a simple profit maximization and no strategic interaction is involved. The first-order conditions corresponding to (3.4) lead to the following set of equations:

$$(3.5) \qquad\qquad PQ_K = r$$

$$PQ_L = W - \frac{(1-\gamma)}{\gamma} \frac{\Pi}{L}$$

where $Q_K = \dfrac{\partial Q}{\partial K}$ = marginal product of capital and $Q_L = \dfrac{\partial Q}{\partial L}$ = marginal

product of labour. Equations (3.5) indicate that while the condition for capital utilization, $PQ_K = r$, is the same as the one for a traditional profit maximizing firm, the condition for labour,

$$PQ_L = W - \left(\frac{1-\gamma}{\gamma}\right) \frac{\Pi}{L},$$

is different. In particular, for $\Pi > 0$ and $\gamma < 1$, the marginal revenue product of labour is equated to a value which is less than W. The implications of this result for the choice of technique are obvious. Conditions (3.5) can be rewritten as,

$$(3.6) \qquad\qquad \frac{Q_K}{Q_L} = \frac{r}{W - (1-\gamma)\Pi} > \frac{r}{W} \text{ if } \Pi > 0 \text{ and } \gamma < 1.$$
$$\frac{}{\gamma} \frac{}{L}$$

The term Q_K/Q_L represents the rate of technical substitution along an isoquant. In the case under consideration, the greater the value of Q_K/Q_L, the lower will be the K/L ratio (more labour-intensive will be the technique) used by the firm. Hence, as can be seen for a given $\Pi > 0$, the technique employed by the firm will be more labour-intensive the greater is $(1-\gamma)$, the influence of the party advocating employment maximization. This result of course, makes sense. Note also that when $\P = 0$, $Q_K/Q_L = r/W$ and the firm uses the same technique as a profit maximizing firm, irrespective of the value of $(1 - \gamma)$. This follows from the fact that when $\P = 0$ the firm has to use the least cost technique it can, given its price environment. This technique is given by $Q_K/Q_L = r/W$. All other techniques are more costly and would presumably lead to the closing of the firm as $\P < 0$. The other party would of course not press this alternative even if its power $(1 - \gamma)$ were high because closure means $L = 0$, which is the worst outcome from the point of view of maximization of L.

The results of the foregoing neoclassical example are depicted graphically in Figures 1 and 2. Condition (3.6) is represented by points A and B in Figure 1. Given the technological options (the production function) and the prices, the optimal K, L allocation of the profit maximizing firm is given by point A. If the party maximizing employment has no power $(1-\gamma = 0)$, the decision-making model yields the same solution, A. If, however, $\pi > 0$ and $1 -\gamma > 0$, the selected technique will be more labour-intensive corresponding for example to point B in Figure 1. In general, for $\pi > 0$ the greater is $(1 -\gamma)$, the more labour-intensive the selected technique. While Figure 1 describes the options in the input (technology) space, Figure 2 shows the possible outcomes in the preference (goal) space. For the example at hand, the frontier giving the maximum combinations of π and L is increasing up to point A then decreasing as L increases beyond L*. This reflects the fact that a profit maximization solution requires some positive amount of input $L = L^*$. In the interval [O, L*] the interests of the two parties are therefore not in direct conflict with one another. It is important to note that the model can accommodate such a situation. It is also crucial that the model be able to eliminate automatically that part of the set for which $L < L^*$, because in that region it is in the interest of both parties to increase L. The assumption of Pareto efficiency (A.2) which is built into the model guarantees that this indeed happens. The same assumption also ensures that, beyond point L*, the set of possible outcomes is given only by the frontier A, L_{MAX}. The profit maximizing firm would select point A. If the L maximizing party has some positive power $(1 - \gamma > 0)$, however, the observed outcome will be at some other point on the frontier, say C. As one moves from A to C, employment is increased at the expense of profit. The outcome at point C reflects a given $0 < \gamma < 1$ and can be characterized as the maximum obtainable value of $\pi^\gamma . L^{(1 -\gamma)}$, which is the objective function derived in this section by equations (3.4) and (3.3).[7]

In concluding the graphical illustration it is important to point out that not all tradeoffs in the preference space lead to tradeoffs in the technology (input) space. For instance, if the variable on the horizontal axis in Figure 2 represented a lump sum payment to another party (recreational facilities to workers or a tax to the government), the profit could be divided among the two parties but there would be no effect on technology as γ varied in the interval [0,1]. It would still be in the interest of the two parties to select the profit maximizing techniques. It is easy to show that the model is constructed so that

Figure 1 The Choice of Production Technique when the Labour Market is Competitive (Not Organized)

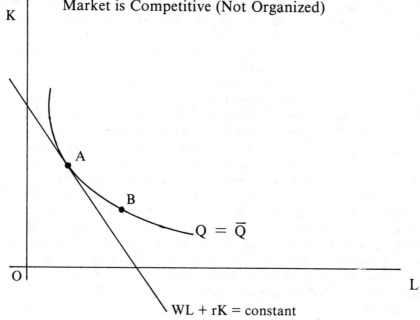

Figure 2 The Choice of Production Technique when the Labour Market is Partly Monopolized (Organized)

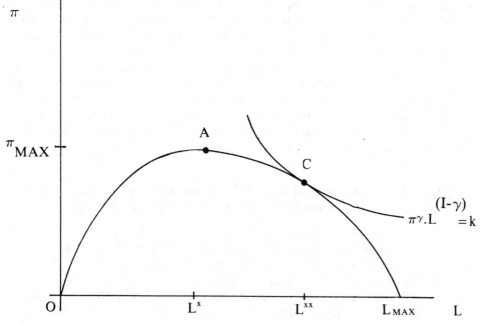

this phenomenon is taken into account.

The most important policy implication of the present approach is that conditions (3.5) can be solved and the resulting reduced forms permit the estimation of $\gamma(\Gamma)$. For a fairly non-technical exposition of these aspects see Svejnar (1980a). A more technical version with econometric estimates from other economic contexts can be found in Svejnar (1980b).

There are several empirical areas of research in which the decision-making model can be fruitfully employed with obvious policy implications. One of these research areas is the choice of technology by multinational corporations (MNCs). These corporations are often viewed as intransigent and utilizing capital-intensive technologies (techniques) even in circumstances where more labour-intensive alternatives exist and possibly are more "appropriate." Our earlier work (interview and questionnaire surveys) suggests that MNCs usually pursue several goals (not just profit maximization) and definitely interact with other agents (governments and trade unions in the host countries, to mention just two), whose professed goals are at least partially different from those of the MNCs. In these cases it would be naturally of interest to see the extent to which these other goals are reflected in the choice of technology (techniques) by the MNCs. If the selected techniques do not reflect these other goals sufficiently, and if alternative techniques exist, and Γ can be altered by public policy, then the findings would have substantial policy implications.

Another area in which the decision-making model can be highly relevant is the choice of technology in agriculture. While some of the best analytical work on technology choices has been done with respect to agriculture, all models assume that it is the individual farmer who solely makes the technological choices. At the same time practitioners acknowledge that extension agents, research and development organizations, government officials, credit organizations and other agents exert considerable influence on the choice of agricultural technology. In view of the importance of (1) agriculture in most LDCs and (2) technology in the agricultural sector, a better understanding of the outcomes of the technological decision-making process in agriculture is essential. In fact, an interdisciplinary team composed of Cornell economists, agricultural specialists and engineers is currently examining the environments in several countries in which the fully developed decision-making model could in the future be applied.

The Social Accounting Matrix as a Data and Conceptual Framework to Explore the Macroeconomic Effects of Alternative Technologies

The Social Accounting Matrix (SAM) provides a useful data and conceptual framework within which, first, technology can be incorporated into a comprehensive macroeconomic model; and, second, the effects of alternative technologies on major policy objectives can be explored. The figures and tables which follow capture and summarize briefly the essence of the SAM and how it can be adapted to deal with the above two questions.

Perhaps the simplest starting point in describing the SAM framework as an analytical system is to recall the interrelationship between: (a) the structure of production; (b) the distribution of the value added, generated by the production activities and yielding the factorial income distribution; and (c) the income distribution by socio-economic groups and the corresponding consumptions and savings behaviour of these socio-economic groups.[8] Figure 3 shows this interrelationship. More specifically, the production activities can be broken down according to types of commodity as well as technology and form of organization, such as for example: (a) furniture produced with essentially labour-intensive techniques in a family enterprise—workshop setting in the informal sector as opposed to furniture being produced in a factory setting using essentially capital intensive technologies; and (b) wheat produced on subsistence farms in the traditional sector as opposed to wheat produced on large-scale farms using mechanized techniques. Hence, the value added which

Figure 3 Simplified Interrelationship Among Principal SAM Accounts (Production Activities, Factors and Institutions)*

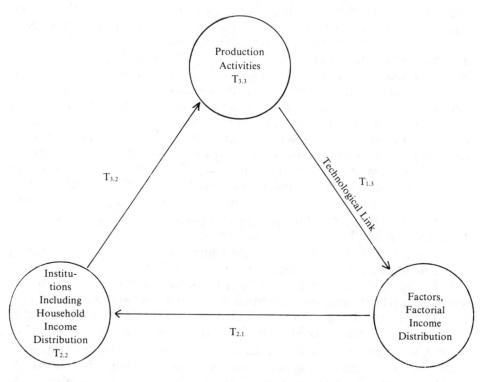

*T stands for the corresponding matrix in the simplified SAM which appears on Table 6. Thus, for example, $T_{1.3}$ refers to the matrix which appears at the intersection of row 1 (account 1), i.e., "factors" and column 3 (account 3), i.e., "production activities".

is generated by the various production activities gets mapped into a factorial income distribution where factors are broken down according to skill or educational levels for labour, capital and land. This is the transformation which is denoted by the matrix $T_{1.3}$ in Figure 3. It is through this transformation that the effects of alternative technologies on the system are captured (i.e. what is called the technological link in Figure 3).

Table 1 represents a basic Social Accounting Matrix (SAM). The major characteristics of a SAM are that it is a square table consisting of as many columns as rows; the expenditures incurred by the accounts, or more properly speaking by the variables into which each account has been broken down, are entered in columns and the receipts are entered in the rows. Since, by definition, the sum of expenditures or outlays of a given subaccount has to be equal to the sum of the receipts of the same subaccount, the total expenditures appearing at the bottom of Table 1 have to be equal to the corresponding total receipts appearing in the last column of Table 1. Perhaps the easiest way to describe the SAM in Table 1 is to focus on the most important transformations which appear in it. Starting with the expenditure side of production activities (Column 6) one can identify the value-added payments to factors (i.e. the matrix at the intersection of Column 6 and Row 1 denoted by the circle 1.a). Likewise, 1.b shows the purchases of intermediate products by the various production activities (i.e. the typical input/output transaction matrix). Next the incomes of the domestic factors of production (i.e. the factorial income distribution) which appears in circle 2 is mainly determined by the distribution of value added (1a). In turn, reading down column 1 labour and non-labour income are allocated to households and companies, respectively. In turn, to obtain the household income distribution, one has to read along Row 2 where the various forms of income accruing to households appear, i.e., the allocation of labour income to households (circle 3), current transfers between households (circle 4a), profit distributed to domestic households (circle 4b) and government transfers to domestic households (circle 4c). Consequently, the household income distribution appears under totals (circle 5). The final transformation which might be highlighted is the consumption expenditures on domestic goods by the various household groups which appear at the intersection of row 6 and column 2 and is denoted by circle 6.

In order to illustrate how the SAM approach can be useful in capturing the technological dimension and as a basis for a model to estimate the effects of different technologies (in fact, different products embodying alternative technologies) on the major development objectives, a SAM table was built for Korea. The underlying data set which was used in the process was that gathered by Adelman and Robinson (1978), in their econometric model of South Korea.[9] Since their model was based on an implicit—rather than explicit— SAM, the task of constructing an actual SAM was accordingly facilitated. On the other hand, the discipline required in actually producing an explicit SAM reveals inconsistencies in the initial data set which have to be explained and reconciled. This proved to be a time-consuming but clarifying activity.

Table 1 A Basic Social Accounting Matrix (SAM)

Receipts		1 Factors of production	2 Households	3 Companies	4 Government	5 Combined capital account	6 Production activities	7 Rest of the world combined account	Totals
			Institutions — Current accounts						
Institutions / Current accounts	1 Factors of production						1ª Value-added payments to factors	Net factor income received from abroad	2 Incomes of the domestic factors of production
	2 Households	3 Allocation of labour income to households	4ª Current transfers between households	4b Profits distributed to domestic households	4c Current transfers to domestic households			Net non-factor incomes received from abroad	5 Incomes of the domestic institutions after transfers
	3 Companies	Allocation of operating surplus to companies			Current transfers to domestic companies				
	4 Government		Direct taxes on income and indirect taxes on current expenditure	Direct taxes on companies plus operating surplus of state enterprise		Indirect taxes on capital goods	Indirect taxes on inputs	Net non-factor incomes received plus indirect taxes on exports	
	5 Combined capital account		Household savings	Undistributed profits after tax	Government current account surplus			Net capital received from abroad	Aggregate savings
	6 Production activities		Household consumption 6 expenditure on domestic goods		Government current expenditure	Investment expenditure on domestic goods	1b Raw material purchases of domestic goods	Exports	Aggregate demand = gross outputs
	7 Rest of the world combined account		Household consumption expenditure on imported goods			Imports of capital goods	Imports of raw materials		Imports
	Totals	Income of the domestic factors of production	Total outlay of households	Total outlay of companies	Total outlay of government	Aggregate investment	Total costs	Total foreign exchange receipts	

Expenditures

Source: Pyatt-Thorbecke (1976), p. 27.

In general, the classification scheme used by Adelman and Robinson (A-R) in their model was also adopted in the present SAM for Korea with one major exception, namely in the taxonomy selected in disaggregating production activities. In fact, given the focus of this study, a crucial question was how to distinguish production activities, at least partially, according to the technology which is embodied in them. The A-R model divided production activities into 29 sectors. Each of these sectors, in turn, was further subdivided into four firm sizes, respectively, according to the number of employees, i.e. large (200 employees and over), medium (50 to 199), small (5 to 49) and self-employed (less than 5). Hence, the A-R model included 116 (29×4) potential activities.[10] It is important to note, at this time, that within a sector the production relationships which were specified in the A-R model for the output of the four firm sizes differed with regard to requirements of primary inputs (capital and labour skills) per unit of output but not with regard to intermediate inputs. In other words, the I-O coefficient matrix ($A_{3.3}$ in Table 6) assumes identical coefficients within a sector for each firm-size. Realistically, however, one should expect significant differences to exist in the use of intermediate inputs by large as opposed to small (or self-employed) firms within, at least, some sectors. For example, the self-employed (family) enterprises producing furniture in workshops in the informal sector are likely to use a very different mix of intermediate inputs than large firms relying on capital-intensive techniques. Unfortunately, data limitations precluded distinguishing the different pattern of intermediate input demand by firm size within a sector. We shall return to this question subsequently.

A set of technological indicators by sectors and firm-size was prepared to help establish a classification of production activities which incorporates at least to some extent, technological differences. Table 2 shows three such indicators, the capital-output ratio, the share of capital value-added to total value-added and the ratio of skilled to unskilled workers. A few relevant observations are suggested by this table: (1) the generally low magnitudes of the capital-output ratios can be partially explained by the use of historical cost pricing for capital while output is valued at current prices; (2) the insignificant differences in the magnitudes of these indicators for the two agricultural sectors (cereals and other) by farm size reflects the uniformity in the adoption of agricultural technology in South Korea and the impact of what has been called a unimodal development strategy in this sector; (3) economies of scale appear to prevail in a few sectors, judging by the strong negative correlation between firm size and the magnitude of the capital-output ratio in processed foods, machinery, and beverages and tobacco;[11] (4) larger firms within a sector tend to employ a higher ratio of skilled to unskilled workers than smaller firms.[12]

The classification of product-cum-technology activities which was ultimately chosen for the Korea-SAM appears in Table 3. It was based mainly on the technological indicators appearing in Table 2 (a few other indicators such as the capital-labour ratio were also looked at). This classification yielded 29

Table 2 Selected Technological Indicators by Sector and Firm-size

Sector Name	Sector Number	Capital/Output Ratio				Value added to Capital = Total Value Added				Ratio of Skilled to Unskilled Workers		
		S	M	L	SE	S	M	L	SE	S	M	L
Cereals	1	.57	.64	.52	.56	.42	.39	.48	.55	–	–	–
Other agriculture	2	.10	.12	.11	.17	.56	.46	.49	.47	–	–	–
Fishery	3	–	2.40	–	2.50	–	.30	–	.63	–	.46	–
Processed foods	4	.94	.44	.23	1.00	.18	.35	.58	.70	.21	.49	.27
Mining	5	1.65	.90	.66	–	.28	.51	.40	.34	.17	.26	.18
Textiles	6	.95	.92	1.00	.79	.42	.46	.54	.44	.13	.16	.10
Finished textiles	7	.21	.30	.22	.21	.53	.32	.31	.75	.04	.11	.09
Lumber/furniture	8,9	1.18	.61	.77	1.22	.43	.57	.65	.55	.10	.37	.23
Chemical intermediates/Cons.	10,11	1.32	1.37	1.33	2.31	.23	.47	.77	.51	.42	.73	.70
Energy[1]	12,13,22	.48	2.29	1.75	.57	.33	.69	.58	.48	.14	.77	1.97
Cement	14	.56	4.15	1.39	.63	.33	.24	.67	.43	.11	.36	.39
Metal products	15	.93	.56	.56	.80	.50	.65	.66	.68	.21	.44	.39
Machinery/electrical machinery	16,17	.68	.35	.28	.73	.43	.51	.47	.66	.27	.40	.48
Transportation equipment	18	.67	.43	.42	.85	.35	.36	.48	.64	.30	.54	.58
Beverages and tobacco	19	1.02	.71	.39	1.75	.21	.46	.81	.62	.24	.58	.27
Other Consumables	20	.58	.40	.41	.60	.46	.54	.51	.74	.24	.25	.21
Construction	21	.21	.21	.21	.21	.23	.21	.47	.50	1.47	.71	.37
Housing	23	30.04	–	–	30.04	.79	–	–	.79	.77	–	.19
Transport and communication	24	4.42	4.42	4.42	4.42	.72	.66	.50	.67	.19	.14	.19
Trade/banking	25	.54	.54	.54	.54	.39	.63	.40	.65	3.14	3.04	5.94
Education	26	–	5.15	–	–	–	.01	–	–	–	5.23	–
Services[2]	27-29	2.32	1.95	2.30	2.30	.30	.24	.46	.38	.17	.49	.26

[1]Petroleum Products (12), Coal (13), Electricity and Water (22)
[2]Medical (27), Other Services (23) and Personal Services (27)
Source: Based on Adelman-Robinson (1978) data set.

SE	= Self employed	(1 to 4 employees)
S	= Small Size Firms	(5 to 49 employees)
M	= Medium Size Firms	(50 to 199 employees)
L	= Large Size Firms	(over 200 employees)

Table 3   Classification of Product-cum-Technology Activities—
Korea SAM

	1	2
Cereals	A	
Other agriculture	A	
Fishery	A	
Processed foods	L	S+SE+M
Mining	A	
Textiles	L	S+SE+M
Finished textiles	A	
Lumber/furniture	A	
Chemical intermediates/ Consumables	L	S+SE+M
Energy	L+M	S+SE
Cement	A	
Metal Products	L+M	S+SE
Machinery/electrical machinery	A	
Transportation equipment	A	
Beverages and tobacco	L	M+S+SE
Other consumables	A	
Construction	A	
Housing	A	
Transportation and communication	A	
Trade/banking	L+M+S	+SE
Education	A	
Services	A	

Total number of production activities: 29

A = Aggregate   L = Large Farm Size   SE = Self employed
                  M = Medium Firm Size
                  S = Small Firm Size

different activities covering 22 sectors. In other words, 7 sectors were broken
down in a dualistic fashion between large scale (using presumably modern
vintage technology) and small scale (using older and, in some instances,

Table 4 Format of the Social Accounting Matrix of South Korea

				Expenditures					
RECEIPTS \ EXPENDITURES		1	2	3	4	5	6	7	
		Factors	Households	Production Activities	Government	Capital	Rest of the World	Totals	
Receipts	1	Factors			Allocation of Labour and Capital Value Added to Factors				Incomes of Domestic Factors
	2	Households	Allocation of Labour and Capital Income to Households					Net Factor and Non-Factor Income Received from Abroad	Incomes of Households
	3	Production Activities		Household Expenditures on Domestic Goods and Services	Inter-mediate Inputs	Government Current Expenditures on Goods and Services	Investment Expenditures on Domestic Goods	Exports	Aggregate Demand = Gross Output
	4	Government	Direct Taxes on Companies and Operating Surplus of State Enterprises	Direct Taxes on Income and Indirect Taxes — Current Transfers	Net Indirect Taxes on Inputs — Subsidies			Net Non-Factor Income Received and Indirect Taxes on Exports	Government Income
	5	Capital	Undistributed Profits after Tax	Households Savings		Government Current Account Surplus		Net Capital Received from Abroad	Aggregate Savings
	6	Rest of the World		Household Imports of Cons. Goods and Comp. Imports	Imports of Inter-mediate Inputs and Non-Comp. Imports		Imports of Capital Goods		Total Imports
	7	Totals	Expenditures = Incomes of Factors	Total Expenditures of Households	Total Cost of Gross Output	Total Expenditures of Government	Aggregate Investment	Total Foreign Exchange Receipts	

Table 5 SAM Table for South Korea, 1968—Matrices of Average Expenditures Propensities for Endogenous Accounts (A_n) and Exogenous Accounts (A_1)[a]

				1 FACTORS OF PRODUCTION				
				Engineers	Technicians	Skilled Workers	Apprentices	Unskilled Workers
				1	2	3	4	5
Receipts	1 Households	Factors of Production	Engineers 1.					
			Technicians 2.					
			Skilled Workers 3.					
			Apprentices 4.					
			Unskilled Workers 5.					
			White-Collar Workers 6.					
			Self-employed in Manufacturing 7.					
			Self-employed in Services 8.					
			Capital 9.					
			Agricultural Labourers 10.					
			Farm Size 1 11.					
			Farm Size 2 12.					
			Farm Size 3 13.					
			Farm Size 4 14.					
			Government Workers 15.					
	2	Wage earners	Engineers 16.	0.529	0.002	0.0	0.002	0.000
			Technicians 17.	0.0	0.827	0.008	0.0	0.007
			Skilled Workers 18.	0.0	0.0	0.879	0.141	0.071
			Apprentices 19.	0.0	0.0	0.0	0.047	0.001
			Unskilled Workers 20.	0.0	0.0	0.027	0.269	0.752
			White-Collar Workers 21.	0.0	0.104	0.009	0.119	0.031
		Self-employed	In Manufacturing 22.	0.0	0.0	0.010	0.065	0.024
			In Services 23.	0.0	0.0	0.043	0.266	0.097
		Agriculture	Capitalist 24.	0.079	0.008	0.005	0.032	0.010
			Agricultural Labourers 25.	0.0	0.0	0.0	0.0	0.0
			Farm Size 1 26.	0.0	0.0	0.0	0.0	0.0
			Farm Size 2 27.	0.0	0.0	0.0	0.0	0.0
			Farm Size 3 28.	0.0	0.0	0.0	0.0	0.0
			Farm Size 4 29.	0.0	0.0	0.0	0.0	0.0
			Government Workers 30.	0.392	0.059	0.018	0.060	0.008
	3	Production Activities	Cereals 31.					
			Other Agriculture 32.					
			Fishing 33.					
			Processed Foods (L) 34.					
			Processed Foods (S + M + Self) 35.					
			Mining 36.					
			Textiles (L) 37.					
			Textiles (S + M + Self) 38.					
			Finished Textile Products 39.					
			Lumber and Furniture 40.					
			Chemical Products (L) 41.					
			Chemical Products (S + M + Self) 42.					
			Energy (L + M) 43.					
			Energy (S + Self) 44.					
			Cement, Non-metallic Mineral Products 45.					
			Metal Products (L + M) 46.					
			Metal Products (S + Self) 47.					
			Machinery 48.					
			Transport Equipment 49.					
			Beverages & Tobacco (L) 50.					
			Beverages & Tobacco (S + M + Self) 51.					
			Other Consumer Products 52.					
			Construction 53.					
			Real Estate 54.					
			Transportation & Communication 55.					
			Trade and Banking (S + M + L) 56.					
			Trade and Banking (Self) 57.					
			Education 58.					
			Medical, Personal & Other Services 59.					
			Sub-total: 1+2+3 60.	1.000	1.000	1.000	1.000	1.000
	4		Government Income 61.	0.0	0.0	0.0	0.0	0.0
	5		Capital Account 62.	0.0	0.0	0.0	0.0	0.0
	6		Rest of the World 63.	0.0	0.0	0.0	0.0	0.0
			Sub-total: 4+5+6 64.	0.0	0.0	0.0	0.0	0.0
			Total Expenditures 65.	19640	23774	86007	12666	10469

[a] A_n is contained in rows 1-59, columns 1-59; A_1 is contained in rows 61-63, columns 1-59. Row 65 gives total expenditures in millions of won for each class and corresponds to column 65 which gives corresponding total incomes. Exogenous expenditures appear in columns 61-64 and are also expressed in millions of won.

[b] Energy = Petroleum and Coal Products and Electricity and Water.

Table 5 (Continued)

				1					
				FACTORS OF PRODUCTION					
White-Collar Workers	Self-employed in Manufacturing	Self-employed in Services	Capital	Agricultural Labourers	Farm Size 1	Farm Size 2	Farm Size 3	Farm Size 4	Government Workers
6	7	8	9	10	11	12	13	14	15
0.002	0.001	0.001	0.002	0.0	0.0	0.0	0.0	0.0	0.002
0.0	0.012	0.012	0.003	0.0	0.0	0.0	0.0	0.0	0.0
0.018	0.021	0.022	0.011	0.0	0.0	0.0	0.0	0.0	0.018
0.0	0.001	0.001	0.0	0.0	0.0	0.0	0.0	0.0	0.0
0.0	0.023	0.023	0.010	0.0	0.0	0.0	0.0	0.0	0.0
0.939	0.017	0.018	0.018	0.0	0.0	0.0	0.0	0.0	0.033
0.004	0.891	0.0	0.032	0.0	0.0	0.0	0.0	0.0	0.011
0.017	0.0	0.888	0.187	0.0	0.0	0.0	0.0	0.0	0.044
0.005	0.002	0.002	0.211	0.0	0.0	0.0	0.0	0.0	0.005
0.0	0.0	0.0	0.002	1.000	0.0	0.0	0.0	0.0	0.0
0.0	0.0	0.0	0.056	0.0	1.000	0.0	0.0	0.0	0.0
0.0	0.0	0.0	0.091	0.0	0.0	1.000	0.0	0.0	0.0
0.0	0.0	0.0	0.133	0.0	0.0	0.0	1.000	0.0	0.0
0.0	0.0	0.0	0.043	0.0	0.0	0.0	0.0	1.000	0.0
0.015	0.032	0.033	0.011	0.0	0.0	0.0	0.0	0.0	0.888
1.000	1.000	1.000	0.810	1.000	1.000	1.000	1.000	1.000	1.000
0.0	0.0	0.0	0.119	0.0	0.0	0.0	0.0	0.0	0.0
0.0	0.0	0.0	0.071	0.0	0.0	0.0	0.0	0.0	0.0
0.0	0.0	0.0	0.0	0.0	0.0	0.0	0.0	0.0	0.0
0.0	0.0	0.0	0.190	0.0	0.0	0.0	0.0	0.0	0.0
14274	7395.9	78372	55875	29132	60709	61179	66680	38986	81720

Table 5 (Continued)

				WAGE EARNERS				
				Engineers	Technicians	Skilled Workers	Apprentices	Unskilled Workers
				16	17	18	19	20
Receipts	1	Factors of Production	Engineers 1.					
			Technicians 2.					
			Skilled Workers 3.					
			Apprentices 4.					
			Unskilled Workers 5.					
			White-Collar Workers 6.					
			Self-employed in Manufacturing 7.					
			Self-employed in Services 8.					
			Capital 9.					
			Agricultural Labourers 10.					
			Farm Size 1 11.					
			Farm Size 2 12.					
			Farm Size 3 13.					
			Farm Size 4 14.					
			Government Workers 15.					
	2 Households	Wage Earners	Engineers 16.					
			Technicians 17.					
			Skilled Workers 18.					
			Apprentices 19.					
			Unskilled Workers 20.					
			White-Collar Workers 21.					
		Self-employed	In Manufacturing 22.					
			In Services 23.					
		Agriculture	Capitalist 24.					
			Agricultural Labourers 25.					
			Farm Size 1 26.					
			Farm Size 2 27.					
			Farm Size 3 28.					
			Farm Size 4 29.					
			Government Workers 30.					
	3	Production Activities	Cereals 31.	0.093	0.127	0.161	0.022	0.197
			Other Agriculture 32.	0.113	0.126	0.124	0.144	0.122
			Fishing 33.	0.026	0.028	0.023	0.023	0.018
			Processed Foods (L) 34.	0.057	0.054	0.050	0.059	0.046
			Processed Foods (S + M + Self) 35.	0.049	0.047	0.043	0.051	0.040
			Mining 36.	0.0	0.0	0.0	0.0	0.0
			Textiles (L) 37.	0.003	0.002	0.002	0.002	0.001
			Textiles (S + M + Self) 38.	0.002	0.001	0.001	0.001	0.001
			Finished Textile Products 39.	0.065	0.052	0.065	0.075	0.078
			Lumber and Furniture 40.	0.003	0.0	0.001	0.002	0.001
			Chemical Products (L) 41.	0.016	0.015	0.014	0.017	0.014
			Chemical Products (S + M + Self) 42.	0.005	0.005	0.004	0.005	0.004
			Energy[b] (L + M) 43.	0.015	0.020	0.021	0.023	0.022
			Energy[b] (S + Self) 44.	0.002	0.003	0.003	0.004	0.003
			Cement, Non-metallic Mineral Products 45.	0.002	0.002	0.002	0.002	0.002
			Metal Products (L + M) 46.	0.002	0.001	0.001	0.002	0.001
			Metal Products (S + Self) 47.	0.001	0.0	0.0	0.001	0.0
			Machinery 48.	0.010	0.004	0.004	0.012	0.005
			Transport Equipment 49.	0.009	0.003	0.003	0.010	0.003
			Beverages & Tobacco (L) 50.	0.024	0.036	0.041	0.050	0.046
			Beverages & Tobacco (S + M + Self) 51.	0.010	0.014	0.016	0.020	0.018
			Other Consumer Products 52.	0.027	0.020	0.018	0.020	0.017
			Construction 53.	0.0	0.0	0.0	0.0	0.0
			Real Estate 54.	0.068	0.062	0.054	0.061	0.047
			Transportation & Communication 55.	0.065	0.079	0.067	0.084	0.055
			Trade and Banking (S + M + L) 56.	0.049	0.046	0.048	0.052	0.050
			Trade and Banking (Self) 57.	0.063	0.060	0.062	0.067	0.064
			Education 58.	0.018	0.024	0.017	0.024	0.003
			Medical, Personal & Other Services 59.	0.081	0.068	0.061	0.075	0.053
			Sub-total: 1+2+3 60.	0.876	0.899	0.905	0.906	0.919
	4		Government Income 61.	0.062	0.047	0.041	0.039	0.042
	5		Capital Account 62.	0.039	0.034	0.034	0.034	0.018
	6		Rest of the World 63.	0.023	0.020	0.020	0.021	0.021
			Sub-total: 4+5+6 64.	0.124	0.101	0.095	0.094	0.081
			Total Expenditures 65.	12320	24949	10131	980.2	96086

Table 5 (Continued)

									2 ENDOGENOUS
			HOUSEHOLDS						
SELF-EMPLOYED			AGRICULTURE						
White-Collar Workers	In Manufacturing	In Services	Capitalist	Agricultural Labourers	Farm Size 1	Farm Size 2	Farm Size 3	Farm Size 4	Government Workers
21	22	23	24	25	26	27	28	29	30
0.117	0.139	0.145	0.117	0.283	0.282	0.282	0.229	0.203	0.122
0.122	0.115	0.120	0.114	0.128	0.128	0.128	0.100	0.108	0.124
0.024	0.024	0.025	0.026	0.021	0.021	0.021	0.016	0.015	0.026
0.053	0.051	0.053	0.054	0.038	0.038	0.038	0.031	0.028	0.059
0.046	0.044	0.046	0.047	0.033	0.033	0.033	0.027	0.024	0.051
0.0	0.0	0.0	0.0	0.0	0.0	0.0	0.0	0.0	0.0
0.003	0.001	0.001	0.002	0.034	0.033	0.033	0.031	0.032	0.001
0.002	0.001	0.001	0.001	0.020	0.019	0.019	0.018	0.019	0.001
0.061	0.054	0.056	0.063	0.012	0.012	0.012	0.011	0.011	0.068
0.004	0.001	0.001	0.001	0.005	0.005	0.005	0.006	0.006	0.002
0.017	0.013	0.013	0.011	0.0	0.0	0.0	0.0	0.0	0.018
0.005	0.004	0.004	0.003	0.0	0.0	0.0	0.0	0.0	0.005
0.017	0.018	0.019	0.018	0.026	0.026	0.026	0.021	0.019	0.018
0.003	0.003	0.003	0.003	0.004	0.004	0.004	0.003	0.003	0.003
0.002	0.001	0.002	0.002	0.002	0.002	0.002	0.002	0.002	0.002
0.002	0.001	0.001	0.001	0.0	0.0	0.0	0.007	0.001	0.001
0.001	0.0	0.0	0.001	0.0	0.0	0.0	0.003	0.001	0.0
0.010	0.004	0.004	0.007	0.0	0.0	0.0	0.030	0.008	0.006
0.0	0.002	0.003	0.006	0.0	0.0	0.0	0.028	0.006	0.0
0.039	0.033	0.035	0.030	0.024	0.024	0.024	0.017	0.018	0.036
0.015	0.013	0.014	0.012	0.010	0.010	0.010	0.007	0.007	0.014
0.019	0.015	0.015	0.020	0.046	0.046	0.046	0.041	0.068	0.019
0.0	0.0	0.0	0.0	0.0	0.0	0.0	0.0	0.0	0.0
0.055	0.051	0.053	0.062	0.013	0.013	0.013	0.008	0.008	0.057
0.070	0.053	0.055	0.070	0.034	0.033	0.033	0.030	0.031	0.059
0.048	0.044	0.046	0.045	0.050	0.050	0.050	0.053	0.046	0.050
0.062	0.057	0.059	0.058	0.065	0.064	0.064	0.068	0.060	0.064
0.009	0.016	0.016	0.019	0.014	0.014	0.014	0.019	0.030	0.014
0.072	0.059	0.062	0.063	0.116	0.116	0.115	0.111	0.133	0.082
0.877	0.817	0.852	0.856	0.978	0.974	0.972	0.918	0.888	0.904
0.053	0.126	0.081	0.069	0.0	0.004	0.004	0.005	0.007	0.041
0.048	0.039	0.048	0.057	0.0	0.0	0.002	0.046	0.080	0.034
0.022	0.018	0.018	0.019	0.022	0.022	0.022	0.032	0.025	0.021
0.123	0.183	0.148	0.144	0.022	0.026	0.028	0.082	0.112	0.096
16343	31427	20619	12253	31648	96241	11703	14775	66195	10039

Table 5   (Continued)

EXPENDITURES

				Cereals	Other Agriculture	Fishing	Processed Foods (L)	Processed Foods (S + M + Self)
				31	32	33	34	35
Receipts	1 Households	Factors of Production	Engineers 1.	0.0	0.0	0.011	0.006	0.004
			Technicians 2.	0.0	0.0	0.015	0.006	0.006
			Skilled Workers 3.	0.0	0.0	0.020	0.031	0.027
			Apprentices 4.	0.0	0.0	0.004	0.003	0.007
			Unskilled Workers 5.	0.0	0.0	0.182	0.016	0.030
			White-Collar Workers 6.	0.0	0.0	0.050	0.014	0.029
			Self-employed in Manufacturing 7.	0.0	0.0	0.041	0.0	0.010
			Self-employed in Services 8.	0.0	0.0	0.0	0.0	0.0
			Capital 9.	0.282	0.291	0.303	0.121	0.084
			Agricultural labourers 10.	0.062	0.038	0.0	0.0	0.0
			Farm Size 1 11.	0.110	0.099	0.0	0.0	0.0
			Farm Size 2 12.	0.121	0.091	0.0	0.0	0.0
			Farm Size 3 13.	0.134	0.096	0.0	0.0	0.0
			Farm Size 4 14.	0.092	0.043	0.0	0.0	0.0
			Government Workers 15.	0.0	0.0	0.0	0.0	0.0
	2	Wage Earners	Engineers 16.					
			Technicians 17.					
			Skilled Workers 18.					
			Apprentices 19.					
			Unskilled Workers 20.					
			White-Collar Workers 21.					
		Self-employed	In Manufacturing 22.					
			In Services 23.					
		Agriculture	Capitalist 24.					
			Agricultural Labourers 25.					
			Farm Size 1 26.					
			Farm Size 2 27.					
			Farm Size 3 28.					
			Farm Size 4 29.					
			Government Workers 30.					
	3	Production Activities	Cereals 31.	0.027	0.102	0.0	0.008	0.008
			Other Agriculture 32.	0.051	0.141	0.016	0.236	0.236
			Fishing 33.	0.0	0.0	0.0	0.066	0.066
			Processed Foods (L) 34.	0.022	0.003	0.005	0.065	0.065
			Processed Foods (S + M + Self) 35.	0.019	0.002	0.004	0.056	0.056
			Mining 36.	0.0	0.001	0.004	0.006	0.006
			Textiles (L) 37.	0.0	0.0	0.001	0.0	0.0
			Textiles (S + M + Self) 38.	0.0	0.0	0.0	0.0	0.0
			Finished Textile Products 39.	0.002	0.003	0.034	0.004	0.004
			Lumber and Furniture 40.	0.0	0.002	0.007	0.002	0.002
			Chemical Products (L) 41.	0.018	0.015	0.002	0.017	0.017
			Chemical Products (S + M + Self) 42.	0.004	0.003	0.0	0.004	0.004
			Energy[h] (L + M) 43.	0.0	0.001	0.036	0.016	0.016
			Energy[h] (S + Self) 44.	0.0	0.0	0.006	0.003	0.003
			Cement, Non-metallic Mineral Products 45.	0.001	0.0	0.0	0.002	0.002
			Metal Products (L + M) 46.	0.0	0.008	0.003	0.004	0.004
			Metal Products (S + Self) 47.	0.0	0.003	0.001	0.001	0.001
			Machinery 48.	0.0	0.0	0.0	0.003	0.003
			Transport Equipment 49.	0.0	0.0	0.006	0.0	0.0
			Beverages & Tobacco (L) 50.	0.0	0.001	0.001	0.004	0.004
			Beverages & Tobacco (S + M + Self) 51.	0.0	0.0	0.0	0.002	0.002
			Other Consumer Products 52.	0.001	0.005	0.002	0.015	0.015
			Construction 53.	0.0	0.0	0.0	0.002	0.002
			Real Estate 54.	0.0	0.0	0.0	0.0	0.0
			Transportation & Communication 55.	0.003	0.006	0.007	0.014	0.014
			Trade and Banking (S + M + L) 56.	0.004	0.009	0.017	0.025	0.025
			Trade and Banking (Self) 57.	0.006	0.012	0.021	0.033	0.033
			Education 58.	0.0	0.0	0.0	0.0	0.0
			Medical, Personal & Other Services 59.	0.012	0.001	0.002	0.012	0.012
			Sub-total: 1+2+3 60.	0.971	0.975	0.803	0.797	0.797
	4		Government Income 61.	0.003	0.001	0.004	0.037	0.037
	5		Capital Account 62.	0.006	0.001	0.075	0.009	0.009
	6		Rest of the World 63.	0.020	0.022	0.117	0.157	0.157
			Sub-total: 4+5+6 64.	0.029	0.025	0.197	0.203	0.203
			Total Expenditures 65.	28506	29588	45409	91980	79544

Table 5   (Continued)

EXPENDITURES									
									3
								PRODUCTION	
Mining	Textiles (L)	Textiles (S+M+Self)	Finished textile products	Lumber and furniture	Chemical Products (L)	Chemical Products (S+M+Self)	Energy[b] (L+M)	Energy[b] (S+Self)	Cement, Non-metallic mineral products
36	37	38	39	40	41	42	43	44	45
0.012	0.004	0.002	0.0	0.002	0.013	0.011	0.011	0.0	0.008
0.028	0.007	0.006	0.002	0.004	0.019	0.010	0.016	0.0	0.009
0.172	0.064	0.069	0.094	0.047	0.033	0.029	0.018	0.017	0.061
0.018	0.008	0.014	0.018	0.010	0.004	0.006	0.001	0.002	0.009
0.070	0.008	0.009	0.015	0.014	0.009	0.029	0.009	0.018	0.023
0.035	0.010	0.020	0.012	0.012	0.032	0.048	0.030	0.010	0.032
0.002	0.0	0.002	0.007	0.010	0.0	0.003	0.0	0.005	0.003
0.0	0.0	0.0	0.0	0.0	0.0	0.0	0.0	0.0	0.0
0.113	0.082	0.059	0.109	0.069	0.125	0.098	0.083	0.116	0.100
0.0	0.0	0.0	0.0	0.0	0.0	0.0	0.0	0.0	0.0
0.0	0.0	0.0	0.0	0.0	0.0	0.0	0.0	0.0	0.0
0.0	0.0	0.0	0.0	0.0	0.0	0.0	0.0	0.0	0.0
0.0	0.0	0.0	0.0	0.0	0.0	0.0	0.0	0.0	0.0
0.0	0.0	0.0	0.0	0.0	0.0	0.0	0.0	0.0	0.0
0.0	0.0	0.0	0.0	0.0	0.0	0.0	0.0	0.0	0.0
0.0	0.0	0.0	0.028	0.001	0.008	0.008	0.0	0.0	0.001
0.058	0.041	0.041	0.007	0.017	0.010	0.010	0.0	0.0	0.003
0.0	0.0	0.0	0.0	0.002	0.001	0.001	0.0	0.0	0.0
0.001	0.004	0.004	0.001	0.002	0.002	0.002	0.0	0.0	0.001
0.001	0.003	0.003	0.0	0.002	0.002	0.002	0.0	0.0	0.001
0.002	0.004	0.003	0.001	0.001	0.016	0.016	0.163	0.163	0.108
0.0	0.130	0.130	0.252	0.001	0.001	0.001	0.0	0.0	0.0
0.0	0.073	0.073	0.141	0.001	0.001	0.001	0.0	0.0	0.0
0.006	0.001	0.001	0.018	0.001	0.001	0.001	0.001	0.001	0.002
0.005	0.002	0.001	0.001	0.066	0.004	0.004	0.0	0.0	0.004
0.023	0.018	0.018	0.004	0.013	0.044	0.044	0.004	0.004	0.007
0.005	0.004	0.004	0.001	0.003	0.010	0.010	0.001	0.001	0.002
0.067	0.020	0.020	0.015	0.006	0.055	0.055	0.070	0.070	0.085
0.010	0.003	0.003	0.002	0.001	0.009	0.009	0.011	0.011	0.013
0.004	0.001	0.001	0.0	0.002	0.011	0.011	0.003	0.003	0.090
0.035	0.002	0.002	0.003	0.007	0.009	0.009	0.005	0.005	0.032
0.012	0.001	0.001	0.001	0.002	0.003	0.003	0.002	0.002	0.011
0.025	0.005	0.005	0.009	0.004	0.004	0.004	0.013	0.013	0.007
0.007	0.001	0.0	0.0	0.0	0.001	0.001	0.003	0.003	0.002
0.005	0.003	0.003	0.004	0.004	0.007	0.007	0.002	0.002	0.006
0.002	0.001	0.001	0.001	0.002	0.003	0.003	0.001	0.001	0.002
0.042	0.008	0.008	0.025	0.011	0.030	0.030	0.009	0.009	0.056
0.006	0.001	0.001	0.003	0.001	0.002	0.002	0.022	0.022	0.002
0.0	0.0	0.0	0.0	0.0	0.0	0.0	0.0	0.0	0.0
0.021	0.009	0.008	0.011	0.011	0.019	0.019	0.084	0.084	0.057
0.027	0.017	0.017	0.046	0.054	0.032	0.032	0.014	0.014	0.030
0.035	0.021	0.021	0.059	0.070	0.041	0.041	0.018	0.018	0.039
0.0	0.0	0.0	0.0	0.0	0.0	0.0	0.0	0.0	0.0
0.017	0.009	0.009	0.012	0.011	0.044	0.044	0.010	0.010	0.020
0.864	0.560	0.560	0.904	0.464	0.602	0.602	0.605	0.605	0.825
0.005	0.054	0.054	0.006	0.008	0.032	0.032	0.0142	0.142	0.030
0.091	0.030	0.030	0.013	0.017	0.093	0.093	0.096	0.096	0.073
0.039	0.356	0.356	0.078	0.509	0.274	0.274	0.158	0.158	0.072
0.135	0.440	0.440	0.096	0.534	0.398	0.398	0.396	0.396	0.175
45027	85835	47934	11486	53207	55635	12738	93449	14595	45462

Table 5: (Continued)

						Metal Products (L+M)	Metal Products (S+Self)	Machinery	Transport Equipment	Beverages & Tobacco (L)
					PROPENSITIES					
					ACTIVITIES					
						46	47	48	49	50
Receipts	1	Factors of Production			Engineers	1. 0.007	0.0	0.009	0.006	0.003
					Technicians	2. 0.011	0.007	0.012	0.015	0.015
					Skilled Workers	3. 0.040	0.043	0.063	0.056	0.019
					Apprentices	4. 0.007	0.012	0.013	0.007	0.005
					Unskilled Workers	5. 0.005	0.008	0.007	0.006	0.008
					White-Collar Workers	6. 0.019	0.014	0.024	0.031	0.013
					Self-employed in Manufacturing	7. 0.0	0.014	0.003	0.001	0.0
					Self-employed in Services	8. 0.0	0.0	0.0	0.0	0.0
					Capital	9. 0.172	0.163	0.118	0.070	0.089
					Agricultural Labourers	10. 0.0	0.0	0.0	0.0	0.0
					Farm Size 1	11. 0.0	0.0	0.0	0.0	0.0
					Farm Size 2	12. 0.0	0.0	0.0	0.0	0.0
					Farm Size 3	13. 0.0	0.0	0.0	0.0	0.0
					Farm Size 4	14. 0.0	0.0	0.0	0.0	0.0
					Government Workers	15. 0.0	0.0	0.0	0.0	0.0
	2	Households	Wage-earners		Engineers	16.				
					Technicians	17.				
					Skilled Workers	18.				
					Apprentices	19.				
					Unskilled Workers	20.				
					White-Collar Workers	21.				
			Self-Employed		In Manufacturing	22.				
					In Services	23.				
			Agriculture		Capitalist	24.				
					Agricultural Labourers	25.				
					Farm Size 1	26.				
					Farm Size 2	27.				
					Farm Sise 3	28.				
					Farm Size 4	29.				
					Government Workers	30.				
	3	Production Activities			Cereals	31. 0.007	0.007	0.0	0.0	0.054
					Other Agriculture	32. 0.010	0.010	0.001	0.004	0.165
					Fishing	33. 0.0	0.0	0.0	0.0	0.0
					Processed Foods (L)	34. 0.004	0.004	0.001	0.001	0.019
					Processed Foods (S + M + Self)	35. 0.004	0.004	0.001	0.001	0.017
					Mining	36. 0.029	0.029	0.005	0.001	0.003
					Textiles (L)	37. 0.0	0.0	0.002	0.0	0.0
					Textiles (S + M + Self)	38. 0.0	0.0	0.001	0.0	0.0
					Finished Textile Products	39. 0.001	0.002	0.001	0.002	0.002
					Lumber and Furniture	40. 0.001	0.001	0.005	0.023	0.005
					Chemical Products (L)	41. 0.008	0.008	0.012	0.011	0.013
					Chemical Products (S + M + Self)	42. 0.002	0.002	0.003	0.003	0.003
					Energy[b] (L + M)	43. 0.043	0.043	0.018	0.018	0.011
					Energy[b] (S + Self)	44. 0.007	0.007	0.003	0.003	0.002
					Cement, Non-metallic Mineral Products	45. 0.008	0.008	0.015	0.006	0.021
					Metal Products (L + M)	46. 0.163	0.163	0.111	0.071	0.016
					Metal Products (S + Self)	47. 0.056	0.056	0.038	0.024	0.006
					Machinery	48. 0.009	0.009	0.079	0.027	0.003
					Transport Equipment	49. 0.001	0.001	0.001	0.221	0.001
					Beverages & Tobacco (L)	50. 0.003	0.003	0.006	0.005	0.057
					Beverages & Tobacco (S + M + Self)	51. 0.001	0.001	0.002	0.002	0.023
					Other Consumer Products	52. 0.005	0.005	0.021	0.027	0.048
					Construction	53. 0.001	0.001	0.002	0.001	0.001
					Real Estate	54. 0.0	0.0	0.0	0.0	0.0
					Transportation & Communication	55. 0.026	0.026	0.021	0.015	0.013
					Trade and Banking (S + M + L)	56. 0.040	0.040	0.039	0.040	0.020
					Trade and Banking (Self)	57. 0.052	0.052	0.051	0.052	0.026
					Education	58. 0.0	0.0	0.0	0.0	0.0
					Medical, Personal & Other Services	59. 0.032	0.032	0.017	0.014	0.018
					Sub-total: 1+2+3	60. 0.775	0.775	0.703	0.763	0.688
	4				Government Income	61. 0.0	0.0	0.029	0.018	0.291
	5				Capital Account	62. 0.018	0.018	0.033	0.018	0.006
	6				Rest of the World	63. 0.207	0.207	0.235	0.200	0.016
					Sub-total: 4+5+6	64. 0.225	0.225	0.297	0.237	0.313
					Total Expenditures	65. 57705	19860	61398	64241	71588

Table 5 (Continued)

Beverages & Tobacco (S+M+Self)	Other Consumer Products	Construction	Real Estate	Transportation and Communication	Trade and Banking (S+M+L)	Trade and Banking (Self)	Education	Medical, Personal and Other Services	
51	52	53	54	55	56	57	58	59	
0.002	0.004	0.047	0.0	0.002	0.0	0.0	0.0	0.002	
0.006	0.008	0.034	0.0	0.006	0.0	0.0	0.0	0.010	
0.025	0.052	0.045	0.001	0.014	0.021	0.0	0.0	0.021	
0.004	0.012	0.003	0.0	0.003	0.001	0.0	0.0	0.002	
0.028	0.011	0.030	0.004	0.193	0.104	0.0	0.032	0.090	
0.023	0.022	0.088	0.006	0.067	0.259	0.0	0.563	0.050	
0.002	0.004	0.0	0.0	0.010	0.0	0.0	0.0	0.0	
0.0	0.0	0.004	0.020	0.0	0.0	0.269	0.179	0.064	
0.051	0.140	0.088	0.587	0.152	0.329	0.444	0.0	0.145	
0.0	0.0	0.0	0.0	0.0	0.0	0.0	0.0	0.0	
0.0	0.0	0.0	0.0	0.0	0.0	0.0	0.0	0.0	
0.0	0.0	0.0	0.0	0.0	0.0	0.0	0.0	0.0	
0.0	0.0	0.0	0.0	0.0	0.0	0.0	0.0	0.0	
0.0	0.0	0.0	0.0	0.0	0.0	0.0	0.0	0.0	
0.0	0.0	0.0	0.0	0.0	0.0	0.0	0.0	0.339	
0.054	0.094	0.014	0.0	0.002	0.0	0.0	0.0	0.0	
0.165	0.027	0.016	0.0	0.0	0.0	0.0	0.0	0.002	
0.005	0.0	0.0	0.0	0.0	0.0	0.0	0.0	0.0	
0.013	0.008	0.002	0.0	0.001	0.002	0.002	0.001	0.001	
0.017	0.007	0.002	0.0	0.001	0.002	0.002	0.001	0.0	
0.003	0.006	0.016	0.0	0.0	0.0	0.0	0.001	0.001	
0.09	0.012	0.0	0.0	0.0	0.0	0.0	0.0	0.001	
0.0	0.007	0.0	0.0	0.0	0.0	0.0	0.0	0.0	
0.002	0.003	0.0	0.0	0.002	0.004	0.004	0.0	0.003	
0.003	0.004	0.070	0.0	0.001	0.001	0.001	0.001	0.002	
0.013	0.021	0.008	0.0	0.005	0.001	0.001	0.002	0.013	
0.003	0.005	0.002	0.0	0.001	0.0	0.0	0.0	0.003	
0.011	0.016	0.009	0.0	0.083	0.008	0.008	0.007	0.024	
0.002	0.003	0.001	0.0	0.013	0.001	0.001	0.001	0.004	
0.021	0.002	0.140	0.0	0.003	0.0	0.0	0.0	0.002	
0.016	0.016	0.055	0.0	0.001	0.004	0.004	0.001	0.001	
0.006	0.006	0.019	0.0	0.0	0.001	0.001	0.0	0.0	
0.003	0.007	0.043	0.0	0.009	0.001	0.001	0.003	0.004	
0.001	0.002	0.0	0.0	0.053	0.002	0.002	0.001	0.001	
0.057	0.010	0.015	0.001	0.006	0.013	0.013	0.005	0.004	
0.023	0.004	0.006	0.0	0.002	0.005	0.005	0.002	0.002	
0.048	0.137	0.019	0.0	0.038	0.016	0.016	0.021	0.050	
0.001	0.002	0.002	0.120	0.005	0.009	0.009	0.082	0.016	
0.0	0.0	0.0	0.0	0.0	0.0	0.0	0.0	0.0	
0.013	0.028	0.029	0.0	0.015	0.040	0.040	0.005	0.011	
0.020	0.034	0.026	0.0	0.018	0.013	0.013	0.005	0.011	
0.026	0.044	0.034	0.0	0.023	0.017	0.017	0.007	0.015	
0.0	0.0	0.0	0.0	0.0	0.0	0.0	0.0	0.001	
0.018	0.016	0.015	0.001	0.010	0.020	0.020	0.012	0.035	
0.688	0.773	0.884	0.740	0.739	0.874	0.874	0.934	0.927	
0.291	0.011	0.005	0.116	0.070	0.050	0.050	0.001	0.038	
0.006	0.014	0.010	0.144	0.127	0.059	0.059	0.043	0.006	
0.016	0.202	0.100	0.0	0.065	0.017	0.017	0.023	0.030	
0.313	0.227	0.115	0.260	0.261	0.126	0.126	0.066	0.073	
28631	17213	25460	59984	16529	14705	19032	53768	24080	

Table 5 (Continued)

		EXOGENOUS EXPENDITURES				
		Government Expenditure	Capital Account	Rest of the World	Sub-total: 4+5+6	Total Income
		4	5	6		
		61	62	63	64	65
Factors of Production	1. Engineers					19639.7
	2. Technicians					23774.2
	3. Skilled Workers					86006.9
	4. Apprentices					12666.4
	5. Unskilled Workers					104686.
	6. White-Collar Workers					142743.
	7. Self-employed in Manufacturing					7395.87
	8. Self-employed in Services					78371.9
	9. Capital					558745.
	10. Agricultural Labourers					29132.3
	11. Farm Size 1					60709.2
	12. Farm Size 2					61179.1
	13. Farm Size 3					66680.4
	14. Farm Size 4					38986.9
	15. Government Workers					81720.0
Households — Wage Earners	16. Engineers			528.512	528.512	12320.0
	17. Technicians			1074.93	1074.93	24948.8
	18. Skilled Workers			4386.12	4386.12	101306.
	19. Apprentices			42.5918	42.5918	980.189
	20. Unskilled Workers			4180.12	4180.12	96085.8
	21. White-Collar Workers			7070.00	7070.00	163433.
Self-employed	22. In Manufacturing			1363.91	1363.91	31426.7
	23. In Services			8932.56	8932.56	206186.
Agriculture	24. Capitalist		.	.687500	.687500	122527.
	25. Agricultural Labourers			1408.96	1408.96	31648.1
	26. Farm Size 1			4284.56	4284.56	96241.2
	27. Farm Size 2			5210.00	5210.00	117026.
	28. Farm Size 3			6599.81	6599.81	147746.
	29. Farm Size 4			2956.84	2956.84	66195.0
	30. Government Workers			4337.19	4337.19	100391.
Production Activities	31. Cereals	708.166	-11724.9	112.000	-10904.7	285056.
	32. Other Agriculture	466.054	771.884	3054.00	4291.94	295880.
	33. Fishing	71.0000	44.8770	4425.00	4540.87	45409.1
	34. Processed Foods (L)	308.284	425.145	6493.86	7227.29	91979.5
	35. Processed Foods (S + M + Self)	266.615	367.681	5616.13	6250.43	79544.1
	36. Mining	198.000	1579.67	8583.00	10360.7	45027.0
	37. Textiles (L)	6.93577	10826.1	11274.8	22107.8	85835.0
	38. Textiles (S + M + Self)	3.87316	6045.65	6296.21	12345.7	47934.0
	39. Finished Textile Products	170.036	1640.50	37136.0	38946.5	114861.
	40. Lumber and Furniture	546.857	1874.58	19319.0	21740.4	53206.5
	41. Chemical Products (L)	1186.97	945.357	675.363	2807.69	55635.2
	42. Chemical Products (S + M + Self)	271.779	216.458	154.637	642.874	12738.4
	43. Energy[b] (L + M)	2737.75	511.489	2650.95	5900.19	93449.0
	44. Energy[b] (S + Self)	427.608	79.8893	414.051	921.548	14595.2
	45. Cement, Non-metallic Mineral Products	131.390	-5476.98	1861.00	-3484.59	45461.5
	46. Metal Products (L + M)	121.150	558.663	2718.39	3398.21	57704.8
	47. Metal Products (S + Self)	41.6968	192.278	935.606	1169.58	19860.4
	48. Machinery	931.883	15425.2	5287.00	21644.1	61398.4
	49. Transport Equipment	807.623	31374.7	405.000	32587.4	64241.2
	50. Beverages & Tobacco (L)	1856.51	3770.56	1883.66	7510.73	71587.6
	51. Beverages & Tobacco (S + M + Self)	742.480	1507.97	753.338	3003.79	28630.9
	52. Other Consumer Products	12616.3	2736.76	43789.0	59142.0	172127.
	53. Construction	7929.00	213775.	8042.00	229746.	254597.
	54. Real Estate	.0	5508.89	.0	5508.89	59983.9
	55. Transportation & Communication	4733.89	599.357	29822.0	35155.2	165290.
	56. Trade and Banking (S + M + L)	1477.16	18554.2	4656.40	24687.8	147053.
	57. Trade and Banking (Self)	1911.84	24014.1	6026.60	31952.5	190323.
	58. Education	32864.1	25.9289	.0	32890.0	53768.1
	59. Medical, Personal & Other Services	94696.1	-15269.1	8577.00	88003.9	240803.
	60. Sub-total: 1+2+3	168231.	310902.	273339.	752472.	564488E+07
4	61. Government Income	.0	.0	72300.0	72300.0	308301.
5	62. Capital Account	136220.	.0	108030.	244250.	432992
6	63. Rest of the World	3850.00	122090.	.0	125940.	433669.
	64. Sub-total: 4+5+6	140070.	122090.	180330.	442490.	.119496E+07
	65. Total Expenditures	308301.	432992.	453669.	.1194960	.6839840EO

informal technology) production.

The format which was followed in building the SAM table for Korea is given in Table 4. This shows the accounts and major transactions within and among accounts which are incorporated in the SAM. It should be noted that the format of the Korea SAM differs slightly from that given in the basic SAM Table 1. In addition to a re-ordering of the accounts, the former does not contain an explicit account for "companies". Instead, the factors account receives the capital value added and distributes to households directly. In addition, among other outlays of the factors accounts are "direct taxes on companies plus operating surplus of government enterprises" which appear as an income to the capital account.

The actual SAM appears in Table 5. It can be seen that the factors account is broken down into 15 categories, i.e. six different labour skills, two types of self-employed, capital, five types of agricultural farmers, and government workers. The classification of households is essentially similar to that of factors. Production activities were divided into 29 classes on the basis of the product-cum-technology characteristics just discussed. The other accounts appearing in the SAM are government, capital, and rest of the world.

Before Table 5 can be properly interpreted, it should be noted that in addition to being a useful comprehensive data framework, the SAM can provide the basis for a conceptual framework and a fixed coefficient and price model. This can be seen by looking at Table 6 which is based on Figure 1. The first step in converting SAM into a model is to decide which accounts are endogenous and which are exogenous. In Table 6 three accounts are assumed to be endogenously determined, i.e. factors, households and production activities while all other accounts are exogenous (i.e. government, capital, and the rest of the world). The three endogenous accounts which appear in Table 6 are the same as those depicted graphically in Figure 1. Likewise, the five endogenous transformations involving these three accounts appear in both Figure 1 and in Table 6. A few examples will suffice to illustrate this point: $T_{1.3}$ is the matrix which allocates the value-added generated by the various production activities into income accruing to the various factors of production and $T_{3.3}$ is the requirement of intermediate inputs or input/output transactions, while $T_{3.2}$ reflects the expenditure pattern of the various institutions including household groups for the different commodities (production activities) which they consume. If a certain number of conditions are met—in particular the existence of excess capacity which would allow prices to remain constant, the framework depicted in Table 6 can be used to estimate the effects of exogenous injections such as government expenditures or exports on the whole system.

In Table 7 the row totals for incomes received by endogenous accounts are given by (the column vector) y_n. Incomes accruing to exogenous accounts are denoted by y_x. It is to be noted that because Table 7 is a SAM, its corresponding row and column totals are equal—column totals for endogenous accounts are given by (the row vector) y'_n, while those for exogenous accounts are given by y'_x.

Table 6 Simplified Schematic Social Accounting Matrix

Receipts \ Expenditures		Endogenous Accounts			Exog.	
		Factors	House-holds	Technology Production Activities	Sum of Other A/cs	Totals
		1	2	3	4	5
Factors	1	0	0	$T_{1.3}$	x_1	y_1
Households	2	$T_{2.1}$	$T_{2.2}$	0	x_2	y_2
Production Activities	3	0	$T_{3.2}$	$T_{3.3}$	x_3	y_3
Sum of other Accounts	4	$1_1'$	$1_2'$	$1_3'$	t	y_x
Totals	5	y_1'	y_2'	y_3'	y_x	

Within the body of the SAM the matrix of all transactions is partitioned by the distinction between endogenous and exogenous accounts. The incomes of the endogenous accounts result partly from expenditures by the endogenous accounts (recorded by the elements T_{nn} of the SAM), and partly from expenditures by the exogenous accounts (recorded by the elements of T_{nx}). The latter are referred to as injections, since they are the incomes injected into the endogenous accounts by the exogenous accounts. If the row sums of T_{nn} are denoted by the vector, n, the row sums of T_{nx} by x, then the fact that y_n represents total incomes for the endogenous accounts implies

$$(4.1) \qquad\qquad y_n = n + x$$

Expenditures by the endogenous accounts are partly the elements of T_{nn} and partly those expenditures which accrue as incomes for the exogenous accounts. The latter are referred to as leakages and are recorded as elements of T_{xn}.

Table 7 Schematic Representation of Endogenous and Exogenous
Accounts in a SAM

Receipts \ Expenditure	Endogenous	Sum	Exogenous	Sum	Totals
Endogenous	T_{nn}	n	Injections T_{nx}	x	y_n
Exogenous	Leakages T_{xn}	l	Residual Balances T_{xx}	t	y_x
Totals	y^l_n		y^l		

Finally, the incomes of the exogenous accounts consist partly of the leakages
from the endogenous accounts and partly of the expenditures by exogenous
accounts which are not injections into the endogenous accounts but rather
accrue as incomes for the exogenous accounts. The latter are represented as
elements of T_{xx} and can be regarded as transfers within the exogenous
accounts.

If row sums for the leakages T_{xn} are denoted by 1, and row sums for the
exogenous account transfers T_{xx} by t, then the exogenous incomes y_x are given
by

$$(4.2) \qquad\qquad y_x = 1 + t$$

At this stage, we can return to Table 5. It was compiled on the assumption
that the accounts for factors, households, and production activities are
endogenous. All other accounts in Table 5, i.e. for government, combined
capital, and the rest of the world are assumed exogenous. The injection matrix
T_{nx} and the transfer matrix T_{xx}[13] appear at the extreme right of Table 5 and are
expressed in millions of won. Here injections comprise government and foreign
current transfers to households, and the demands on production activities
resulting from government current expenditure, investment and exports. The
other transaction matrices T_{nn} and T_{xn} do not appear explicitly in Table 5.
Instead two matrices of endogenous expenditures propensities A_n and A_l
appear. The former is square and is analogous to T_{nn}. However, while elements
of T_{nn} are money flows, elements of A_n represent average propensities to
consume. These can be obtained from T_{nn} simply by dividing a particular
element of T_{nn} by the total income for the column account in which the element

occurs. Hence columns of A_n show expenditures as proportions of total income (i.e. y'_n in Table 7) and not as absolute amounts. Similarly, A_1 is analogous to T_{xn} and is derived from it by expressing leakage expenditures as fractions of total income (i.e. y'_x).[14]

Thus, for example, Table 5 shows that 9.3 per cent of expenditure by households headed by engineers was spent on cereals (see row 31 and column 16) and 5.7 per cent on processed foods produced by large firms (row 34, column 16). The corresponding figures for agricultural labourers are 28.3 per cent (row 31, column 25) and 3.8 per cent (row 34, column 25), respectively. Likewise, it can be seen from Table 5 that out of factor reward (i.e. expenditures) for engineering skills, 52.9 per cent accrued to households headed by an engineer (row 16, column 1), 7.9 per cent went to capitalist households (row 24, column 1) and 39.2 per cent to government workers' households (row 31, column 1) reflecting the fact that many of the latter are engineers.

The implications of introducing the matrices A_n and A_1 of average expenditure propensities include the fact that n and l defined in equations (4.1) and (4.2) can now be written as:

(4.3) $$n = A_n \, y_n, \text{ and}$$

(4.4) $$l = A_1 \, y_n$$

Equation (4.3) states that row sums of T_{nn} can be obtained by multiplying the average expenditure propensities for each row of the endogenous accounts by the level of income recorded in each column. Equation (4.4) similarly sums up the effects of endogenous incomes on leakages.

(4.5) $$y_n = A_n y_n + x$$

(4.6) $$= M_a x , \qquad \text{where}$$

(4.7) $$M_a = (I - A_n)^{-1}$$

Thus from (4.6) endogenous incomes can be accounted for by multiplying injections x by a multiplier matrix M_a. This multiplier matrix is to be referred to as the accounting multiplier matrix because it explains the results observed in a SAM and not the process by which they are generated. The process which actually generates a SAM over time would require the specification of a model including a set of dynamic behavioural and technical relationships connecting the different accounts and variables included in the SAM. Such an attempt at obtaining *ex ante* multipliers may be undertaken in a next phase of this project but for the present, we shall rely on accounting multipliers to provide some estimates of the effects of alternative product-cum-technologies on major development objectives.

In this connection, it is important to analyze further the endogenous transactions matrix T_{nn} and its corresponding coefficient matrix A_n. It is really for this purpose that Table 6 was prepared. It is analogous to Table 5 except that: (a) it breaks down explicitly the way in which T_{nn} is partitioned; and (b) it collapses all exogenous accounts into a vector of sums of rows and columns, respectively. Thus, the injections are denoted by x_i where the subscript i stands for the three endogenous accounts (factors, households and production activities) and leakages by l'_i.

Thus, corresponding to the partitioning of T_{nn} in Table 7,

$$(4.8) \qquad A_n = \begin{bmatrix} O & O & A_{1.3} \\ A_{2.1} & A_{2.2} & O \\ O & A_{3.2} & A_{3.3} \end{bmatrix}$$

Hence equation (4.6) above, i.e.

$$(4.6) \qquad y_n = (I - A_n)^{-1}x = M_a x$$

can be interpreted as follows, i.e. the income levels of factors (y_1), households (y_2) and production activities (y_3) are endogenously determined as functions of the injections (the exogenous demand of the other accounts). All the behavioural and technical coefficients of the underlying interdependent system (reflected by a model) are explicitly incorporated in the partitioned (fixed coefficient) matrix A_n. Thus, by way of illustration, $A_{1.3}$ allocates the value added generated by the various production activities to the various factors such as labour skills as a proportion of the value of gross output of each activity (sector). Likewise, $A_{3.3}$ represents the intermediate (input-output) demand. As such, the elements of $A_{1.3}$ and $A_{3.3}$ must be based on an emprical knowledge of the sectoral production functions. Each column of the production activities account represents, in fact, a linear Leontif-type sectoral production function.

Another example should suffice to illustrate the fact that the partitioned A_n matrix contains all the necessary behavioural and technical relations to close the system in a consistent way. $A_{3.2}$ reflects the consumption behaviour of the different socio-economic household groups and other institutions. More specifically, it shows the proportion of incomes (= expenditures) of each household and institutional class which is spent on each production activity. It can readily be verified that the five component matrices of the partitioned matrix A_n appear in Table 5 reflecting the respective average propensities obtained in Korea in 1968.

At first glance, the model specified in equation (4.6) appears analogous to the open Leontief model. The basic difference, however, is that in contrast with the latter, the SAM model is closed with respect to the determination of the factorial and household income distribution and the consumption behaviour of households. Indeed, the open Leontief model can be written as follows using the same notation as in (4.8)

$$(4.9) \qquad y_3 = A_{3.3}y_3 \quad + (A_{3.2}y_2 + x_3)$$

$$= [1 - A_{3.3}]^{-1}(A_{3.2}y_2 + x_3)$$

It allows the vector of total output of production activities (y_3) to be determined endogenously by pre-multiplying the exogenously given vector of final demand (x_3) by the Leontief-inverse (where $A_{3.3}$ is the matrix of input-output coefficients). It is clear that the partitioned matrix A in equation (4.6) incorporates four other matrices in addition to $A_{3.3}$ which permits a much higher degree of closure of the interdependent system.

The accounting multiplier matrix M_a derived in equation (4.6) is reproduced on Table 8. It should be interpreted as follows: An injection of 100 units (i.e. S. Korean wons) into any particular column will typically be responsible for some part of every endogenous income. More specifically, we are interested in the effects of production activities—incorporating given technologies—on the major policy objectives. Thus, for example, it can be asked what the microeconomic consequences would be for an injection of 100 units of exogenous demand for any given product-cum-technology activity. Consider, for example, an injection of 100 units of exogenous demand for mining products (activity 36). This exogenous demand might be through exports or government expenditures. As a result of this injection circulating around the economy, it gives rise to both direct and indirect effects. The column for mining products shows what the gross effects are. They account for 2.3 units of engineers' labour income, 4.6 units of technicians' labour income, 24.6 units of skilled labour income and so on reading down that column[15]. Going further down the same column 36, the implications of an additional exogenous demand for mining products on the income distribution by socio-economic groups can be gathered. Thus, an injection of 100 units additional demand for mining products would yield an additional 1.4 units of income to engineering households (row 16, column 36), 4.4 units to technicians' households (row 17, column 36), 24.8 units to skilled workers' households (row 18, column 36) and so on. Likewise Table 8 can be used to derive some implications regarding a number of aggregate multipliers; thus it can be seen by adding the first 15 entries of column 36 that the value-added multiplier for mining products amounts to 2.0. Similarly adding the next 15 entries (representing the 15 household classes) yields the income multiplier, i.e. 1.853. Finally, the aggregate output multiplier (rows 31 - 59 of column 36) amounts to 4.156. In the process of circulating around the economy, the 100-unit injection ultimately leaks out, as can be seen by looking at the last three elements of column 36. Some 30.9 units leak out as direct taxes on output, 32.8 units contribute to savings and 36.3 units leak out as imports. Total leakages equal the initial injection.

Table 8 can be used to explore a large number of macroeconomic effects of alternative production-cum-technology patterns. Two examples may suffice to illustrate the type of questions which can be answered with the SAM approach. Policy-makers might be interested in the alternative consequences of using different techniques to produce a similar output—such as: (1) a traditional labour-intensive technique used by small scale traditional farmers in wheat production in contrast with a capital-intensive technique relying on tractors and intermediate inputs used by large, commercial farmers; or (2) an informal labour-intensive technique compared to a modern vintage technique in the construction industry. In our Korean-SAM application 7 sectors have been broken down dualistically, on the basis of a number of technological indicators, i.e. processed food, textiles, chemical products, energy, metal products, beverages and tobacco, and trade and banking. By consulting the appropriate multipliers in Table 8 the alternative effects of the two techniques in each of these sectors can be determined. Thus, for instance, it can be seen that an exogenous demand for mining products benefits the incomes of unskilled workers' households significantly more than a corresponding increase for textiles produced by large firms. A 100-unit increase in output of mining products would yield 15.5 units of income to unskilled workers' households (row 20, column 36) in contrast with only 6.1 units if the additional output had been produced by modern textile firms (row 20, column 37).

In this fashion the implications of a given output-mix or technological choice on such policy objectives as aggregate output (or income or value-added), income distribution by socio-economic groups, employment[16] and the balance of payments can be explored.

A second type of question which can be answered by Table 8 is which product-cum-technology activity is likely to benefit target socio-economic groups. For instance, the government might want to know which activities have the greatest multiplier impact on the incomes of unskilled workers, white collar workers or agricultural labourers. This information is readily provided by reading along the relevant rows under production activities. Thus, it can be seen by reading along row 20—which represents income accruing to unskilled workers' households—that the latter benefit most income-wise from an exogenous increase in Trade and Banking output (L + M + S) which yields a multiplier of 1.65, Trade and Banking (self-employed) (.892), Education (.818), Real Estate, Construction (.812) and "Other Consumer Products" (.789). Incomes of white collar households (row 21), in turn, are affected most by Trade and Banking (SE) and, of course, agricultural labourers' incomes respond to changes in agricultural production. However, it is interesting to note that this group also benefits from a number of other production activities which use agricultural inputs such as processed foods (see row 25, columns 34-35). This illustrates the way the SAM multiplier captures the independence within the economy. Even though the production of processed foods requires only little, if any, agricultural labour, directly, it requires agricultural products as inputs which, in turn, tend to be very labour-intensive. Consequently, the indirect effects of increased food processing can be very significant on agricultural labourers' incomes.

Three points should be made by way of concluding this section. First, it is essential to recall that the Korean SAM application in this section was undertaken for demonstrational and illustrative purposes. Clearly the type and degree of disaggregation of production activities which is desirable to analyze the effects of alternative technologies on policy objectives might have to be more thorough than the breakdown of 29 activities adopted in the present treatment. More specifically, a more precise distinction between formal and

Table 8 Accounting Multiplier Matrices, South Korea, 1968[a]

					FACTORS OF PRODUCTION				
					Engineers	Technicians	Skilled Workers	Apprentices	Unskilled Workers
					1	2	3	4	5
1	Production activities		Engineers	1.	1.010	0.010	0.010	0.010	0.010
			Technicians	2.	0.018	1.018	0.019	0.017	0.018
			Skilled Workers	3.	0.079	0.078	1.079	0.078	0.080
			Apprentices	4.	0.012	0.012	0.012	1.012	0.013
			Unskilled Workers	5.	0.124	0.127	0.124	0.121	1.122
			White-Collar Workers	6.	0.138	0.143	0.139	0.136	0.137
			Self-employed in Manufacturing	7.	0.010	0.010	0.010	0.009	0.010
			Self-employed in Services	8.	0.095	0.095	0.095	0.093	0.095
			Capital	9.	0.789	0.802	0.812	0.796	0.821
			Agricultural labourers	10.	0.053	0.056	0.059	0.057	0.061
			Farm Size 1	11.	0.111	0.116	0.121	0.118	0.125
			Farm Size 2	12.	0.112	0.117	0.123	0.120	0.127
			Farm Size 3	13.	0.122	0.128	0.134	0.130	0.138
			Farm Size 4	14.	0.071	0.075	0.079	0.077	0.082
			Government Workers	15.	0.094	0.092	0.090	0.089	0.090
2	Households	Wage-earners	Engineers	16.	0.536	0.009	0.008	0.009	0.007
			Technicians	17.	0.020	0.847	0.028	0.020	0.027
			Skilled Workers	18.	0.095	0.095	0.975	0.235	0.167
			Apprentices	19.	0.001	0.001	0.001	0.048	0.002
			Unskilled Workers	20.	0.109	0.111	0.136	0.376	0.860
			White-Collar Workers	21.	0.157	0.265	0.167	0.273	0.186
		Self-Employed	In Manufacturing	22.	0.040	0.041	0.051	0.104	0.064
			In Services	23.	0.257	0.260	0.304	0.522	0.359
		Agriculture	Capitalist	24.	0.250	0.182	0.180	0.204	0.187
			Agricultural labourers	25.	0.055	0.057	0.060	0.059	0.062
			Farm Size 1	26.	0.155	0.161	0.167	0.163	0.171
			Farm Size 2	27.	0.183	0.190	0.196	0.192	0.201
			Farm Size 3	28.	0.227	0.235	0.242	0.236	0.248
			Farm Size 4	29.	0.106	0.110	0.114	0.111	0.118
			Government Workers	30.	0.498	0.163	0.121	0.162	0.110
3	Factors of production		Cereals	31.	0.525	0.555	0.598	0.578	0.629
			Other Agriculture	32.	0.537	0.555	0.559	0.550	0.563
			Fishing	33.	0.084	0.086	0.082	0.081	0.079
			Processed foods (L)	34.	0.174	0.173	0.171	0.169	0.169
			Processed Foods (S + M + Self)	35.	0.151	0.150	0.148	0.147	0.147
			Mining	36.	0.034	0.035	0.035	0.034	0.035
			Textiles (L)	37.	0.085	0.082	0.086	0.085	0.089
			Textiles (S + M + Self)	38.	0.048	0.047	0.049	0.048	0.051
			Finished textile products	39.	0.149	0.138	0.149	0.148	0.158
			Lumber and furniture	40.	0.018	0.017	0.017	0.017	0.017
			Chemical products (L)	41.	0.074	0.073	0.073	0.072	0.073
			Chemical products (S + M + Self)	42.	0.019	0.019	0.019	0.019	0.019
			Energy[b] (L + M)	43.	0.119	0.125	0.126	0.123	0.127
			Energy[b] (S + Self)	44.	0.019	0.020	0.020	0.019	0.020
			Cement, non-metallic Min Prods.	45.	0.020	0.020	0.020	0.020	0.020
			Metal products (L + M)	46.	0.039	0.038	0.038	0.037	0.038
			Metal products (S + Self)	47.	0.014	0.013	0.013	0.013	0.013
			Machinery	48.	0.039	0.036	0.036	0.036	0.036
			Transport equipment	49.	0.037	0.035	0.035	0.034	0.035
			Beverages & Tobacco (L)	50.	0.102	0.111	0.116	0.114	0.120
			Beverages & Tobacco (S + M + Self)	51.	0.041	0.044	0.046	0.046	0.048
			Other consumer products	52.	0.155	0.154	0.153	0.150	0.153
			Construction	53.	0.033	0.034	0.032	0.032	0.031
			Real Estate	54.	0.125	0.123	0.116	0.114	0.111
			Transportation & Communication	55.	0.208	0.224	0.215	0.206	0.207
			Trade and Banking (S + M + L)	56.	0.191	0.190	0.193	0.190	0.196
			Trade and Banking (Self)	57.	0.247	0.246	0.250	0.247	0.254
			Education	58.	0.044	0.050	0.045	0.042	0.040
			Medical, personal & other services	59.	0.277	0.271	0.266	0.263	0.264
4			Government Income	61.	0.365	0.365	0.364	0.374	0.372
5			Capital Account	62.	0.280	0.282	0.281	0.277	0.271
6			Rest of the World	63.	0.355	0.353	0.355	0.350	0.358

Devised for Table 5. The multiplier matrix M_a is obtained from equation 4.6 in text, i.e. $y_n = (I - A_n)^{-1} y_n + x = M_a y^1_n + x$.
The matrix M_a appears in rows 1-59 and columns 1-59.
Energy = Petroleum and Coal Products and Electricity and Water.

Table 8 (Continued)

White-collar Workers	Self-employed in Manufacturing	Self-employed in Services	Capital	Agricultural Labourers	Farm Size 1	Farm Size 2	Farm Size 3	Farm Size 4	Government Workers
6	7	8	9	10	11	12	13	14	15
0.010	0.010	0.010	0.008	0.011	0.011	0.011	0.010	0.010	0.010
0.017	0.016	0.017	0.014	0.019	0.019	0.019	0.019	0.018	0.018
0.077	0.072	0.075	0.063	0.083	0.083	0.083	0.082	0.078	0.080
0.012	0.011	0.012	0.010	0.013	0.013	0.013	0.013	0.012	0.013
0.123	0.114	0.118	0.099	0.132	0.131	0.131	0.123	0.122	0.125
1.133	0.128	0.133	0.113	0.150	0.149	0.149	0.146	0.147	0.139
0.010	1.009	0.009	0.009	0.010	0.010	0.010	0.009	0.009	0.010
0.092	0.088	0.091	0.078	0.106	0.106	0.106	0.102	0.101	0.096
0.782	0.747	0.776	1.649	0.895	0.892	0.890	0.818	0.793	0.806
0.054	0.054	0.056	0.048	1.073	0.072	0.072	0.063	0.061	0.056
0.113	0.111	0.115	0.099	0.147	1.147	0.147	0.128	0.124	0.116
0.114	0.112	0.117	0.100	0.150	0.150	1.149	0.131	0.127	0.117
0.124	0.122	0.127	0.109	0.164	0.164	0.163	1.143	0.138	0.128
0.073	0.072	0.075	0.065	0.099	0.098	0.098	0.086	1.083	0.075
0.091	0.084	0.088	0.080	0.119	0.119	0.119	0.111	0.117	1.096
0.009	0.007	0.008	0.007	0.008	0.008	0.008	0.007	0.007	0.009
0.019	0.030	0.031	0.019	0.022	0.022	0.021	0.021	0.020	0.020
0.111	0.109	0.113	0.087	0.102	0.101	0.101	0.098	0.095	0.114
0.001	0.002	0.002	0.001	0.001	0.001	0.001	0.001	0.001	0.001
0.108	0.123	0.127	0.097	0.116	0.116	0.116	0.108	0.107	0.110
1.090	0.163	0.169	0.146	0.171	0.171	0.170	0.165	0.166	0.191
0.044	0.929	0.039	0.064	0.044	0.044	0.044	0.040	0.039	0.051
0.270	0.241	1.138	0.397	0.289	0.288	0.287	0.270	0.264	0.305
0.174	0.163	0.169	0.351	0.193	0.193	0.192	0.177	0.171	0.179
0.056	0.055	0.057	0.051	1.074	0.074	0.074	0.065	0.062	0.058
0.156	0.153	0.159	0.191	0.197	1.197	0.196	0.174	0.169	0.161
0.185	0.180	0.187	0.250	0.232	0.231	1.230	0.205	0.198	0.190
0.228	0.222	0.231	0.329	0.284	0.283	0.282	1.252	0.244	0.235
0.107	0.105	0.109	0.136	0.138	0.137	0.137	0.121	1.117	0.110
0.118	0.128	0.132	0.100	0.131	0.130	0.130	0.122	0.126	0.996
0.536	0.542	0.562	0.499	0.786	0.783	0.781	0.680	0.647	0.554
0.541	0.518	0.538	0.442	0.614	0.612	0.610	0.536	0.535	0.557
0.081	0.079	0.082	0.065	0.086	0.086	0.085	0.075	0.072	0.045
0.168	0.162	0.168	0.134	0.174	0.174	0.173	0.154	0.149	0.178
0.146	0.140	0.146	0.116	0.151	0.150	0.150	0.134	0.129	0.154
0.034	0.032	0.033	0.028	0.038	0.038	0.038	0.036	0.034	0.034
0.083	0.077	0.080	0.076	0.119	0.119	0.119	0.110	0.109	0.096
0.047	0.044	0.045	0.043	0.068	0.068	0.068	0.063	0.062	0.048
0.142	0.133	0.138	0.101	0.105	0.105	0.105	0.098	0.096	0.151
0.020	0.016	0.017	0.016	0.023	0.023	0.023	0.023	0.022	0.018
0.073	0.067	0.069	0.054	0.067	0.067	0.056	0.061	0.060	0.076
0.019	0.017	0.018	0.014	0.016	0.016	0.016	0.015	0.015	0.020
0.120	0.115	0.119	0.101	0.141	0.140	0.140	0.128	0.123	0.123
0.019	0.018	0.019	0.016	0.022	0.022	0.022	0.020	0.019	0.019
0.020	0.018	0.019	0.016	0.021	0.021	0.021	0.020	0.019	0.020
0.038	0.034	0.036	0.032	0.039	0.038	0.038	0.052	0.039	0.038
0.013	0.012	0.012	0.011	0.013	0.013	0.013	0.018	0.014	0.013
0.040	0.032	0.034	0.032	0.033	0.033	0.033	0.064	0.039	0.037
0.031	0.031	0.033	0.032	0.033	0.033	0.032	0.065	0.037	0.031
0.112	0.103	0.107	0.083	0.107	0.107	0.107	0.094	0.094	0.111
0.045	0.041	0.043	0.033	0.043	0.043	0.043	0.038	0.038	0.045
0.150	0.139	0.144	0.134	0.200	0.200	0.199	0.183	0.211	0.154
0.031	0.030	0.031	0.025	0.031	0.031	0.030	0.029	0.029	0.033
0.115	0.109	0.113	0.081	0.083	0.083	0.082	0.073	0.072	0.119
0.213	0.191	0.198	0.161	0.199	0.199	0.198	0.185	0.182	0.208
0.189	0.178	0.185	0.156	0.211	0.210	0.210	0.203	0.192	0.194
0.244	0.230	0.239	0.201	0.273	0.272	0.271	0.263	0.249	0.251
0.037	0.042	0.044	0.038	0.047	0.047	0.047	0.050	0.060	0.043
0.268	0.249	0.258	0.235	0.352	0.350	0.349	0.327	0.344	0.283
0.364	0.411	0.382	0.409	0.340	0.343	0.342	0.318	0.313	0.362
0.286	0.264	0.281	0.299	0.261	0.260	0.262	0.288	0.315	0.280
0.350	0.326	0.338	0.292	0.399	0.398	0.397	0.395	0.373	0.358

Table 8   (Continued)

				WAGE EARNERS				
				Engineers	Technicians	Skilled Workers	Apprentices	Unskilled Workers
				16	17	18	19	20
1	Factors of Production		Engineers	1.	0.010	0.010	0.010	0.011	0.010
			Technicians	2.	0.018	1.018	0.018	0.018	0.018
			Skilled Workers	3.	0.079	0.078	0.079	0.082	0.081
			Apprentices	4.	0.012	0.012	0.012	0.013	0.013
			Unskilled Workers	5.	0.123	0.128	0.125	0.128	0.123
			White-Collar Workers	6.	0.138	0.144	0.140	0.145	0.137
			Self-employed in Manufacturing	7.	0.010	0.010	0.010	0.010	0.010
			Self-employed in Services	8.	0.094	0.096	0.095	0.096	0.095
			Capital	9.	0.777	0.804	0.815	0.780	0.833
			Agricultural labourers	10.	0.051	0.056	0.059	0.049	0.062
			Farm Size 1	11.	0.107	0.117	0.122	0.104	0.128
			Farm Size 2	12.	0.108	0.118	0.123	0.104	0.130
			Farm Size 3	13.	0.118	0.128	0.135	0.113	0.142
			Farm Size 4	14.	0.069	0.075	0.079	0.064	0.084
			Government Workers	15.	0.093	0.092	0.090	0.091	0.090
2	Households	Wage Earners	Engineers	16.	1.007	0.007	0.007	0.007	0.007
			Technicians	17.	0.020	1.020	0.020	0.020	0.020
			Skilled Workers	18.	0.095	0.095	1.096	0.098	0.098
			Apprentices	19.	0.001	0.001	0.001	1.001	0.001
			Unskilled Workers	20.	0.108	0.112	0.110	0.112	1.109
			White-Collar Workers	21.	0.157	0.163	0.159	0.164	0.157
		Self-employed	In Manufacturing	22.	0.040	0.041	0.041	0.040	0.041
			In Services	23.	0.254	0.261	0.262	0.257	0.266
		Agriculture	Capitalist	24.	0.168	0.174	0.176	0.169	0.180
			Agricultural labourers	25.	0.053	0.058	0.061	0.050	0.064
			Farm Size 1	26.	0.151	0.162	0.167	0.147	0.175
			Farm Size 2	27.	0.179	0.191	0.197	0.174	0.205
			Farm Size 3	28.	0.221	0.236	0.243	0.217	0.253
			Farm Size 4	29.	0.103	0.110	0.115	0.098	0.120
			Government Workers	30.	0.103	0.105	0.103	0.104	0.103
3	Production Activities		Cereals	31.	0.503	0.558	0.601	0.435	0.650
			Other agriculture	32.	0.524	0.557	0.561	0.555	0.569
			Fishing	33.	0.084	0.087	0.082	0.080	0.078
			Processed foods (L)	34.	0.171	0.174	0.171	0.173	0.169
			Processed foods (S + M + Self)	35.	0.148	0.151	0.148	0.150	0.147
			Mining	36.	0.033	0.035	0.035	0.036	0.036
			Textiles (L)	37.	0.085	0.082	0.086	0.086	0.092
			Textiles (S + M + Self)	38.	0.048	0.045	0.049	0.049	0.052
			Finished textile products	39.	0.147	0.137	0.150	0.158	0.164
			Lumber and furniture	40.	0.019	0.017	0.017	0.019	0.017
			Chemical products (L)	41.	0.073	0.073	0.073	0.073	0.074
			Chemical products (S + M + Self)	42.	0.019	0.019	0.019	0.019	0.019
			Energy[b] (L + M)	43.	0.117	0.126	0.127	0.129	0.129
			Energy[b] (S + Self)	44.	0.018	0.020	0.020	0.020	0.020
			Cement, non-metallic min. products	45.	0.019	0.020	0.020	0.021	0.020
			Metal products (L + M)	46.	0.039	0.037	0.038	0.042	0.039
			Metal products (S + Self)	47.	0.014	0.013	0.013	0.015	0.013
			Machinery	48.	0.041	0.035	0.036	0.044	0.036
			Transport equipment	49.	0.041	0.036	0.035	0.044	0.035
			Beverages & tobacco (L)	50.	0.096	0.111	0.117	0.125	0.124
			Beverages & tobacco (S + M + Self)	51.	0.038	0.044	0.047	0.050	0.049
			Other consumer products	52.	0.157	0.154	0.154	0.153	0.155
			Construction	53.	0.034	0.034	0.033	0.034	0.031
			Real Estate	54.	0.129	0.124	0.117	0.122	0.110
			Transportation & communication	55.	0.208	0.189	0.216	0.231	0.207
			Trade and banking (S + M + L)	56.	0.189	0.190	0.194	0.194	0.199
			Trade and banking (Self)	57.	0.244	0.246	0.251	0.251	0.258
			Education	58.	0.045	0.053	0.046	0.051	0.039
			Medical, personal & other services	59.	0.275	0.271	0.267	0.268	0.265
4			Government Income	61.	0.368	0.366	0.362	0.361	0.370
5			Capital Account	62.	0.278	0.282	0.281	0.279	0.267
6			Rest of the World	63.	0.354	0.353	0.356	0.360	0.363

Table 8 . (Continued)

									ENDOGENOUS
					2				
			HOUSEHOLDS						
	SELF-EMPLOYED					AGRICULTURE			
White-collar Workers	In Manufacturing	In Services	Capitalists	Agricultural Labourers	Farm Size 1	Farm Size 2	Farm Size 3	Farm Size 4	Government Workers
21	22	23	24	25	26	27	28	29	30
0.010	0.009	0.010	0.010	0.011	0.011	0.011	0.010	0.010	0.010
0.017	0.016	0.017	0.017	0.019	0.019	0.019	0.019	0.018	0.018
0.077	0.071	0.074	0.076	0.083	0.083	0.083	0.082	0.078	0.080
0.012	0.011	0.012	0.012	0.013	0.013	0.013	0.013	0.012	0.013
0.123	0.113	0.117	0.120	0.132	0.131	0.131	0.123	0.122	0.126
0.133	0.127	0.133	0.135	0.150	0.149	0.149	0.146	0.147	0.140
0.010	0.009	0.009	0.010	0.010	0.010	0.010	0.009	0.009	0.010
0.092	0.087	0.091	0.090	0.106	0.106	0.106	0.102	0.101	0.096
0.782	0.740	0.772	0.765	0.895	0.892	0.890	0.818	0.793	0.809
0.054	0.053	0.055	0.053	0.073	0.072	0.072	0.063	0.061	0.056
0.112	0.110	0.115	0.109	0.147	0.147	0.147	0.128	0.124	0.116
0.113	0.111	0.116	0.110	0.150	0.150	0.149	0.131	0.127	0.117
0.124	0.121	0.127	0.120	0.164	0.164	0.163	0.143	0.138	0.129
0.072	0.072	0.075	0.071	0.099	0.098	0.098	0.086	0.083	0.075
0.091	0.083	0.087	0.086	0.119	0.119	0.119	0.111	0.117	0.097
0.007	0.007	0.007	0.007	0.008	0.008	0.008	0.007	0.007	0.007
0.019	0.018	0.019	0.019	0.022	0.022	0.021	0.021	0.020	0.020
0.093	0.086	0.090	0.091	0.102	0.101	0.101	0.098	0.095	0.097
0.001	0.001	0.001	0.001	0.001	0.001	0.001	0.001	0.001	0.001
0.108	0.099	0.103	0.105	0.116	0.116	0.116	0.108	0.107	0.111
1.151	0.144	0.151	0.153	0.171	0.171	0.170	0.165	0.166	0.159
0.039	0.037	0.039	0.039	0.044	0.044	0.044	0.040	0.039	0.041
0.252	0.238	1.248	0.247	0.289	0.288	0.287	0.270	0.264	0.262
0.169	0.160	0.167	1.165	0.193	0.193	0.192	0.177	0.171	0.175
0.056	0.055	0.057	0.054	1.074	0.074	0.074	0.065	0.062	0.058
0.156	0.151	0.158	0.152	0.197	1.197	0.196	0.174	0.169	0.162
0.184	0.178	0.186	0.180	0.232	0.231	1.230	0.205	0.198	0.190
0.228	0.220	0.230	0.222	0.284	0.283	0.282	1.252	0.252	0.236
0.106	0.104	0.108	0.104	0.138	0.137	0.137	0.121	1.117	0.110
0.103	0.095	0.099	0.098	0.131	0.130	0.130	0.122	0.126	1.109
0.535	0.537	0.560	0.524	0.786	0.783	0.781	0.680	0.647	0.554
0.541	0.514	0.536	0.522	0.614	0.612	0.610	0.536	0.535	0.559
0.081	0.078	0.082	0.082	0.086	0.086	0.085	0.075	0.072	0.085
0.168	0.161	0.168	0.168	0.174	0.174	0.173	0.154	0.149	0.180
0.146	0.139	0.145	0.145	0.151	0.150	0.150	0.134	0.129	0.156
0.033	0.031	0.033	0.033	0.038	0.038	0.038	0.038	0.034	0.035
0.083	0.076	0.079	0.082	0.119	0.119	0.119	0.110	0.109	0.086
0.047	0.043	0.045	0.046	0.068	0.068	0.068	0.063	0.062	0.049
0.142	0.131	0.136	0.144	0.105	0.105	0.105	0.098	0.096	0.153
0.020	0.016	0.017	0.016	0.023	0.023	0.023	0.023	0.022	0.018
0.073	0.066	0.069	0.066	0.067	0.067	0.066	0.061	0.060	0.076
0.019	0.017	0.018	0.017	0.016	0.016	0.016	0.015	0.015	0.020
0.120	0.114	0.118	0.118	0.141	0.140	0.140	0.128	0.123	0.123
0.019	0.018	0.019	0.018	0.022	0.022	0.022	0.020	0.019	0.019
0.020	0.018	0.019	0.019	0.021	0.021	0.021	0.020	0.019	0.020
0.039	0.034	0.035	0.037	0.039	0.038	0.038	0.052	0.039	0.038
0.014	0.012	0.012	0.013	0.013	0.013	0.013	0.018	0.014	0.013
0.040	0.032	0.033	0.035	0.033	0.033	0.033	0.064	0.039	0.037
0.030	0.031	0.033	0.037	0.033	0.033	0.032	0.065	0.037	0.031
0.112	0.102	0.106	0.101	0.107	0.107	0.107	0.094	0.094	0.112
0.045	0.041	0.043	0.040	0.043	0.043	0.043	0.038	0.038	0.045
0.150	0.137	0.143	0.147	0.200	0.200	0.199	0.183	0.211	0.155
0.031	0.030	0.031	0.032	0.031	0.031	0.030	0.029	0.029	0.033
0.115	0.108	0.112	0.121	0.083	0.083	0.082	0.073	0.072	0.120
0.214	0.188	0.196	0.211	0.199	0.199	0.198	0.185	0.182	0.208
0.189	0.176	0.184	0.182	0.211	0.210	0.210	0.203	0.192	0.195
0.244	0.228	0.238	0.236	0.273	0.272	0.271	0.263	0.249	0.253
0.037	0.042	0.044	0.046	0.047	0.047	0.047	0.050	0.060	0.043
0.269	0.246	0.256	0.254	0.352	0.350	0.349	0.327	0.344	0.286
0.364	0.416	0.384	0.370	0.340	0.343	0.342	0.318	0.313	0.360
0.287	0.262	0.281	0.292	0.261	0.260	0.262	0.288	0.315	0.280
0.350	0.322	0.336	0.339	0.399	0.398	0.397	0.395	0.373	0.360

Table 8 (Continued)

				Cereals 31	Other Agriculture 32	Fishing 33	Processed Foods (L) 34	Processed Foods (S+M+Self) 35
1	Factors of Production		Engineers	1.	0.010	0.010	0.019	0.016	0.013
			Technicians	2.	0.017	0.017	0.029	0.022	0.022
			Skilled Workers	3.	0.074	0.073	0.084	0.096	0.092
			Apprentices	4.	0.012	0.011	0.014	0.013	0.017
			Unskilled Workers	5.	0.115	0.113	0.273	0.122	0.137
			White-Collar Workers	6.	0.132	0.131	0.154	0.125	0.140
			Self-employed in Manufacturing	7.	0.009	0.008	0.048	0.010	0.020
			Self-employed in Services	8.	0.092	0.092	0.072	0.074	0.075
			Capital	9.	1.065	1.119	0.894	0.824	0.792
			Agricultural labourers	10.	0.124	0.108	0.041	0.052	0.052
			Farm Size 1	11.	0.237	0.243	0.084	0.112	0.113
			Farm Size 2	12.	0.249	0.236	0.085	0.111	0.112
			Farm Size 3	13.	0.275	0.255	0.093	0.121	0.122
			Farm Size 4	14.	0.175	0.137	0.055	0.068	0.068
			Government Workers	15.	0.101	0.095	0.065	0.068	0.068
2	Households	Wage Earners	Engineers	16.	0.007	0.007	0.012	0.010	0.009
			Technicians	17.	0.020	0.020	0.031	0.023	0.024
			Skilled Workers	18.	0.093	0.093	0.112	0.109	0.108
			Apprentices	19.	0.001	0.001	0.001	0.001	0.001
			Unskilled Workers	20.	0.103	0.103	0.223	0.108	0.120
			White-Collar Workers	21.	0.156	0.155	0.179	0.144	0.159
		Self-employed	In Manufacturing	22.	0.048	0.049	0.081	0.041	0.050
			In Services	23.	0.305	0.314	0.270	0.244	0.242
		Agriculture	Capitalist	24.	0.028	0.240	0.195	0.178	0.172
			Agricultural labourers	25.	0.126	0.110	0.043	0.054	0.054
			Farm Size 1	26.	0.296	0.305	0.134	0.159	0.157
			Farm Size 2	27.	0.346	0.338	0.166	0.186	0.184
			Farm Size 3	28.	0.417	0.404	0.212	0.231	0.227
			Farm Size 4	29.	0.221	0.186	0.094	0.104	0.103
			Government Workers	30.	0.115	0.110	0.088	0.085	0.085
3	Production Activities		Cereals	31.	1.638	0.711	0.411	0.429	0.432
			Other Agriculture	32.	0.569	1.661	0.395	0.659	0.662
			Fishing	33.	0.073	0.070	1.055	0.124	0.124
			Processed foods (L)	34.	0.170	0.149	0.119	1.178	0.179
			Processed foods (S + M + Self)	35.	0.147	0.129	0.103	0.154	1.155
			Mining	36.	0.033	0.034	0.036	0.036	0.037
			Textiles (L)	37.	0.094	0.092	0.073	0.062	0.062
			Textiles (S + M + Self)	38.	0.054	0.053	0.041	0.035	0.035
			Finished textile products	39.	0.099	0.100	0.128	0.086	0.087
			Lumber and furniture	40.	0.019	0.021	0.020	0.016	0.016
			Chemical products (L)	41.	0.079	0.078	0.051	0.070	0.071
			Chemical products (S + M + Self)	42.	0.019	0.019	0.013	0.017	0.017
			Energy[b] (L + M)	43.	0.118	0.117	0.124	0.110	0.110
			Energy[b] (S + Self)	44.	0.018	0.018	0.020	0.017	0.017
			Cement, non-metallic mineral products	45.	0.019	0.018	0.015	0.017	0.017
			Metal products (L + M)	46.	0.038	0.048	0.033	0.037	0.037
			Metal products (S + Self)	47.	0.013	0.017	0.011	0.013	0.013
			Machinery	48.	0.037	0.036	0.028	0.030	0.030
			Transport equipment	49.	0.036	0.036	0.034	0.027	0.027
			Beverages & tobacco (L)	50.	0.090	0.090	0.077	0.072	0.073
			Beverages & tobacco (S + M + Self)	51.	0.035	0.036	0.031	0.029	0.029
			Other consumer products	52.	0.166	0.166	0.114	0.132	0.133
			Construction	53.	0.027	0.027	0.023	0.023	0.023
			Real Estate	54.	0.075	0.076	0.072	0.060	0.061
			Transportation & communication	55.	0.177	0.179	0.153	0.150	0.152
			Trade and Banking (S + M + L)	56.	0.183	0.186	0.152	0.159	0.160
			Trade and Banking (Self)	57.	0.237	0.241	0.197	0.205	0.207
			Education	58.	0.042	0.041	0.029	0.028	0.028
			Medical, Personal & other services	59.	0.298	0.279	0.192	0.199	0.200
4			Government Income	61.	0.346	0.350	0.308	0.316	0.314
5			Capital Account	62.	0.279	0.276	0.301	0.229	0.228
6			Rest of the World	63.	0.375	0.374	0.392	0.456	0.358

Table 8 (Continued)

EXPENDITURES									
									3
									PRODUCTION
Mining	Textiles (L)	Textiles (S+M+Self)	Finished textile products	Lumber and Furniture	Chemical Products (L)	Chemical Products (S+M+Self)	Energy[b] (L+M)	Energy[b] (S+Self)	Cement, Non-metallic mineral products
36	37	38	39	40	41	42	43	44	45
0.023	0.010	0.008	0.009	0.006	0.020	0.019	0.021	0.010	0.020
0.046	0.017	0.017	0.018	0.012	0.031	0.023	0.033	0.017	0.029
0.246	0.117	0.122	0.182	0.084	0.085	0.082	0.096	0.094	0.155
0.029	0.016	0.022	0.032	0.016	0.012	0.014	0.011	0.012	0.023
0.173	0.064	0.066	0.104	0.070	0.084	0.105	0.102	0.110	0.136
0.154	0.076	0.087	0.120	0.081	0.118	0.134	0.119	0.098	0.150
0.010	0.004	0.006	0.013	0.014	0.005	0.008	0.006	0.011	0.011
0.081	0.044	0.044	0.076	0.055	0.060	0.060	0.051	0.051	0.074
0.781	0.454	0.435	0.699	0.430	0.572	0.549	0.521	0.549	0.709
0.044	0.024	0.024	0.036	0.021	0.027	0.027	0.025	0.025	0.035
0.091	0.051	0.051	0.075	0.044	0.056	0.056	0.052	0.052	0.072
0.092	0.051	0.051	0.076	0.044	0.056	0.057	0.052	0.052	0.073
0.100	0.055	0.056	0.083	0.048	0.062	0.062	0.057	0.057	0.079
0.058	0.032	0.032	0.049	0.028	0.036	0.037	0.034	0.034	0.047
0.073	0.039	0.039	0.060	0.038	0.060	0.060	0.046	0.046	0.066
0.014	0.006	0.005	0.007	0.004	0.012	0.011	0.012	0.007	0.012
0.044	0.017	0.017	0.020	0.013	0.013	0.023	0.031	0.018	0.030
0.248	0.118	0.124	0.185	0.089	0.094	0.092	0.103	0.102	0.163
0.002	0.001	0.001	0.002	0.001	0.001	0.001	0.001	0.001	0.001
0.155	0.061	0.064	0.100	0.065	0.076	0.092	0.088	0.095	0.122
0.179	0.088	0.099	0.139	0.092	0.132	0.147	0.132	0.112	0.169
0.043	0.023	0.025	0.042	0.031	0.028	0.030	0.027	0.032	0.040
0.259	0.143	0.141	0.229	0.147	0.179	0.178	0.164	0.169	0.230
0.172	0.099	0.095	0.152	0.093	0.125	0.120	0.114	0.119	0.155
0.045	0.025	0.025	0.038	0.022	0.028	0.028	0.026	0.026	0.036
0.135	0.076	0.075	0.114	0.068	0.088	0.087	0.081	0.082	0.112
0.163	0.092	0.091	0.139	0.083	0.108	0.107	0.100	0.102	0.137
0.204	0.116	0.114	0.176	0.105	0.138	0.135	0.127	0.130	0.174
0.092	0.052	0.051	0.079	0.047	0.061	0.060	0.056	0.057	0.078
0.098	0.051	0.051	0.077	0.048	0.076	0.075	0.064	0.058	0.086
0.423	0.224	0.225	0.370	0.206	0.273	0.276	0.251	0.251	0.351
0.452	0.262	0.264	0.344	0.210	0.259	0.262	0.245	0.242	0.335
0.055	0.020	0.030	0.045	0.030	0.036	0.036	0.034	0.033	0.046
0.117	0.065	0.066	0.097	0.060	0.076	0.077	0.071	0.070	0.099
0.101	0.057	0.057	0.084	0.052	0.066	0.067	0.062	0.061	0.085
1.047	0.025	0.025	0.030	0.017	0.050	0.050	0.201	0.201	0.167
0.064	1.196	0.196	0.373	0.032	0.041	0.042	0.038	0.038	0.054
0.036	0.110	1.110	0.209	0.018	0.023	0.024	0.022	0.022	0.031
0.102	0.050	0.051	1.094	0.047	0.060	0.062	0.060	0.058	0.083
0.020	0.010	0.010	0.013	1.078	0.014	0.014	0.012	0.012	0.019
0.078	0.051	0.052	0.056	0.040	1.082	0.082	0.041	0.041	0.058
0.019	0.012	0.013	0.014	0.010	0.020	1.020	0.010	0.010	0.014
0.173	0.078	0.079	0.105	0.056	0.130	0.130	1.157	0.156	0.201
0.027	0.012	0.012	0.016	0.009	0.020	0.020	0.024	1.024	0.011
0.022	0.010	0.010	0.014	0.011	0.024	0.024	0.018	0.018	1.115
0.082	0.021	0.021	0.033	0.026	0.036	0.036	0.041	0.041	0.083
0.029	0.007	0.007	0.012	0.009	0.013	0.013	0.014	0.014	0.029
0.058	0.022	0.022	0.036	0.018	0.026	0.026	0.040	0.040	0.040
0.037	0.016	0.016	0.023	0.015	0.020	0.020	0.029	0.029	0.032
0.084	0.044	0.045	0.069	0.044	0.058	0.059	0.052	0.051	0.075
0.034	0.018	0.018	0.027	0.018	0.023	0.023	0.021	0.020	0.030
0.168	0.075	0.075	0.131	0.074	0.117	0.118	0.097	0.097	0.182
0.032	0.014	0.014	0.024	0.013	0.020	0.020	0.040	0.040	0.026
0.074	0.037	0.038	0.058	0.036	0.046	0.047	0.045	0.044	0.061
0.181	0.092	0.093	0.141	0.090	0.125	0.126	0.193	0.191	0.208
0.172	0.095	0.096	0.169	0.128	0.126	0.127	0.109	0.108	0.161
0.223	0.123	0.124	0.219	0.166	0.164	0.165	0.142	0.140	0.208
0.030	0.016	0.016	0.025	0.015	0.019	0.019	0.018	0.018	0.025
0.216	0.116	0.116	0.177	0.111	0.175	0.176	0.136	0.134	0.194
0.309	0.231	0.230	0.284	0.167	0.245	0.244	0.350	0.351	0.313
0.328	0.164	0.163	0.221	0.140	0.262	0.261	0.277	0.277	0.310
0.363	0.606	0.607	0.495	0.692	0.493	0.495	0.374	0.372	0.377

Table 8 (Continued)

				PROPENSITIES				
				ACTIVITIES				
				Metal Products (L+M)	Metal Products (S+Self)	Machinery	Transport Equipment	Beverages & Tobacco (L)
				46	47	48	49	50
1	Factors of Production		Engineers	1.	0.016	0.010	0.018	0.016	0.011
			Technicians	2.	0.029	0.025	0.027	0.034	0.018
			Skilled Workers	3.	0.113	0.116	0.129	0.134	0.077
			Apprentices	4.	0.019	0.023	0.024	0.019	0.014
			Unskilled Workers	5.	0.099	0.103	0.090	0.092	0.088
			White-Collar Workers	6.	0.128	0.123	0.123	0.140	0.105
			Self-employed in Manufacturing	7.	0.007	0.021	0.010	0.008	0.006
			Self-employed in Services	8.	0.076	0.076	0.069	0.073	0.063
			Capital	9.	0.761	0.753	0.647	0.624	0.678
			Agricultural labourers	10.	0.033	0.033	0.029	0.030	0.045
			Farm Size 1	11.	0.069	0.069	0.060	0.061	0.096
			Farm Size 2	12.	0.070	0.070	0.061	0.062	0.096
			Farm Size 3	13.	0.076	0.076	0.067	0.068	0.105
			Farm Size 4	14.	0.045	0.045	0.039	0.040	0.060
			Government Workers	15.	0.068	0.068	0.057	0.057	0.061
2	Households	Wage Earners	Engineers	16.	0.010	0.007	0.011	0.010	0.007
			Technicians	17.	0.029	0.026	0.027	0.033	0.019
			Skilled Workers	18.	0.123	0.126	0.135	0.139	0.088
			Apprentices	19.	0.001	0.001	0.001	0.001	0.001
			Unskilled Workers	20.	0.092	0.096	0.086	0.086	0.081
			White-Collar Workers	21.	0.147	0.143	0.140	0.156	0.121
		Self-employed	In Manufacturing	22.	0.037	0.049	0.035	0.033	0.032
			In Services	23.	0.234	0.234	0.207	0.206	0.203
		Agriculture	Capitalist	24.	0.165	0.163	0.141	0.136	0.146
			Agricultural labourers	25.	0.035	0.035	0.030	0.031	0.047
			Farm Size 1	26.	0.112	0.111	0.097	0.096	0.134
			Farm Size 2	27.	0.139	0.138	0.120	0.119	0.158
			Farm Size 3	28.	0.178	0.177	0.153	0.151	0.195
			Farm Size 4	29.	0.078	0.078	0.067	0.067	0.090
			Government Workers	30.	0.086	0.084	0.075	0.075	0.073
3	Production Activities		Cereals	31.	0.337	0.338	0.297	0.300	0.412
			Other Agriculture	32.	0.322	0.322	0.279	0.287	0.513
			Fishing	33.	0.044	0.044	0.040	0.040	0.044
			Processed foods (L)	34.	0.097	0.097	0.085	0.086	0.112
			Processed foods (S + M + Self)	35.	0.084	0.084	0.074	0.075	0.097
			Mining	36.	0.072	0.072	0.040	0.034	0.031
			Textiles (L)	37.	0.051	0.051	0.048	0.047	0.051
			Textiles (S + M + Self)	38.	0.029	0.029	0.027	0.026	0.029
			Finished textile products	39.	0.075	0.076	0.069	0.072	0.067
			Lumber and furniture	40.	0.014	0.014	0.017	0.042	0.018
			Chemical products (L)	41.	0.054	0.054	0.053	0.055	0.059
			Chemical products (S + M + Self)	42.	0.013	0.013	0.013	0.014	0.014
			Energy[b] (L + M)	43.	0.142	0.142	0.104	0.107	0.091
			Energy[b] (S + Self)	44.	0.022	0.022	0.016	0.017	0.014
			Cement, non-metallic mineral products	45.	0.025	0.025	0.031	0.022	0.037
			Metal products (L + M)	46.	1.239	0.239	0.170	0.147	0.051
			Metal products (S + Self)	47.	0.082	1.082	0.062	0.051	0.018
			Machinery	48.	0.037	0.037	1.110	0.061	0.027
			Transport equipment	49.	0.026	0.026	0.023	1.305	0.023
			Beverages & Tobacco (L)	50.	0.066	0.066	0.064	0.066	1.119
			Beverages & Tobacco (S + M + Self)	51.	0.026	0.027	0.026	0.026	0.048
			Other consumer products	52.	0.108	0.108	0.117	0.131	0.158
			Construction	53.	0.022	0.022	0.021	0.020	0.019
			Real Estate	54.	0.057	0.057	0.053	0.053	0.050
			Transportation & Communication	55.	0.161	0.161	0.143	0.142	0.130
			Trade and Banking (S + M + L)	56.	0.165	0.165	0.153	0.165	0.133
			Trade and Banking (Self)	57.	0.213	0.213	0.198	0.213	0.173
			Education	58.	0.024	0.024	0.022	0.022	0.023
			Medical, personal & other services	59.	0.201	0.201	0.168	0.168	0.178
4			Government Income	61.	0.265	0.265	0.265	0.258	0.544
5			Capital Account	62.	0.230	0.230	0.220	0.208	0.188
6			Rest of the World	63.	0.506	0.506	0.516	0.533	0.269

Table 8 (Continued)

Beverages & Tobacco (S+M+Self) 51	Other Consumer Products 52	Construction 53	Real Estate 54	Transportation and Communication 55	Trade and Banking (S+M+L) 56	Trade and Banking (Self) 57	Education 58	Medical, Personal and Other Services 59
0.010	0.012	0.173	0.169	0.168	0.168	0.174	0.173	0.149
0.020	0.023	0.035	0.034	0.034	0.033	0.038	0.038	0.034
0.083	0.119	0.047	0.048	0.047	0.045	0.068	0.068	0.062
0.013	0.023	0.017	0.017	0.020	0.017	0.023	0.023	0.022
0.110	0.104	0.019	0.019	0.019	0.018	0.016	0.016	0.015
0.117	0.129	0.020	0.019	0.019	0.019	0.022	0.022	0.019
0.008	0.011	0.038	0.037	0.038	0.036	0.039	0.038	0.039
0.064	0.074	0.036	0.036	0.040	0.034	0.033	0.033	0.039
0.645	0.756	0.111	0.114	0.112	0.107	0.107	0.107	0.094
0.046	0.043	0.154	0.150	0.150	0.144	0.200	0.199	0.211
0.097	0.088	0.123	0.114	0.115	0.113	0.083	0.082	0.072
0.097	0.090	0.190	0.190	0.189	0.185	0.211	0.210	0.192
0.106	0.099	0.050	0.042	0.037	0.044	0.047	0.047	0.060
0.061	0.059	0.365	0.374	0.364	0.382	0.340	0.342	0.313
0.061	0.000	0.353	0.350	0.350	0.338	0.399	0.397	0.373
0.006	1.010	0.010	0.010	0.010	0.008	0.011	0.010	0.010
0.021	0.079	1.079	0.080	0.072	0.063	0.083	0.082	0.080
0.095	0.124	0.124	1.122	0.114	0.099	0.131	0.123	0.125
0.001	0.010	0.010	0.010	1.009	0.008	0.010	0.009	0.010
0.097	0.798	0.812	0.821	0.747	1.649	0.892	0.818	0.806
0.132	0.111	0.121	0.125	0.111	0.099	1.147	0.128	0.116
0.033	0.122	0.134	0.138	0.122	0.109	0.164	1.143	0.128
0.200	0.094	0.090	0.090	0.084	0.080	0.119	0.111	1.096
0.140	0.020	0.028	0.027	0.030	0.019	0.022	0.021	0.020
0.047	0.001	0.001	0.002	0.002	0.001	0.001	0.001	0.001
0.133	0.157	0.167	0.186	0.163	0.146	0.171	0.165	0.191
0.156	0.257	0.304	0.359	0.241	0.397	0.288	0.270	0.305
0.192	0.055	0.060	0.062	0.055	0.051	0.074	0.065	0.058
0.089	0.183	0.196	0.201	0.180	0.250	0.231	0.205	0.190
0.074	0.106	0.114	0.118	0.105	0.136	0.137	0.121	0.110
0.416	0.525	0.598	0.629	0.542	0.499	0.783	0.680	0.554
0.518	0.084	0.082	0.079	0.079	0.065	0.086	0.075	0.085
0.045	0.151	0.148	0.147	0.140	0.116	0.150	0.134	0.154
0.113	0.085	0.086	0.089	0.077	0.076	0.119	0.110	0.086
0.098	0.149	0.149	0.158	0.133	0.101	0.105	0.098	0.151
0.032	0.074	0.073	0.073	0.067	0.054	0.067	0.061	0.076
0.052	0.119	0.126	0.127	0.115	0.101	0.140	0.128	0.123
0.029	0.020	0.020	0.020	0.018	0.016	0.021	0.020	0.020
0.069	0.014	0.013	0.013	0.012	0.011	0.013	0.018	0.013
0.018	0.037	0.035	0.035	0.031	0.032	0.033	0.065	0.031
0.059	0.041	0.046	0.048	0.041	0.033	0.043	0.038	0.045
0.014	0.033	0.032	0.031	0.030	0.025	0.031	0.029	0.033
0.092	0.208	0.215	0.207	0.191	0.161	0.199	0.185	0.208
0.014	0.247	0.250	0.254	0.230	0.201	0.272	0.263	0.251
0.037	0.277	0.266	0.264	0.249	0.235	0.350	0.327	0.283
0.051	0.280	0.281	0.271	0.264	0.299	0.260	0.288	0.280
0.018	0.000	0.000	0.000	0.000	0.000	0.000	0.000	0.000
0.027	1.018	0.017	0.017	0.017	0.019	0.019	0.018	0.018
0.023	0.012	1.012	0.012	0.012	0.013	0.013	0.012	0.012
0.120	0.143	0.136	1.133	0.133	0.150	0.149	0.147	0.138
1.048	0.095	0.093	0.092	1.091	0.106	0.106	0.101	0.094
0.158	0.056	0.057	0.054	0.056	1.073	0.072	0.061	0.051
0.019	0.117	0.120	0.114	0.117	0.150	1.149	0.127	0.108
0.051	0.075	0.077	0.073	0.075	0.099	0.098	1.083	0.069
0.132	0.009	0.009	0.009	0.008	0.008	0.008	0.007	1.007
0.135	0.095	0.235	0.111	0.113	0.102	0.101	0.095	0.095
0.175	0.111	0.376	0.108	0.127	0.116	0.116	0.107	0.108
0.023	0.041	0.104	0.044	0.039	0.044	0.044	0.039	0.040
0.180	0.182	0.204	0.174	0.169	0.193	0.192	0.171	0.168
0.543	0.288	0.290	0.311	0.304	0.323	0.323	0.313	0.316
0.187	0.420	0.421	0.454	0.442	0.464	0.464	0.453	0.464
0.271	0.292	0.289	0.235	0.254	0.214	0.214	0.235	0.220

informal types of technology (and form of organization) in production, given reasonably homogenous commodities, is called for. In particular, it is important to obtain reliable estimates of the production functions faced by different producers and of the actual techniques selected by them. Information is needed both with regard to primary inputs (e.g. labour skill groups, and capital) and to intermediate inputs. It should be recalled that the data base used in building the SAM production columns for Korea assumed the same intermediate input pattern for all firm sizes. Such a distinction is essential if one wants to obtain a better understanding of the degree of interdependence between formal and informal production activities. This is an area of work which we expect to pursue in the future.

A second point is that the use of accounting multipliers to trace through the effects of alternative technologies on the whole economic system is based on a set of restrictive simplifying assumptions. Instead of relying on accounting multipliers—which presume constant average expenditures propensities[17], it is fairly easy, at least for some transactions to substitute marginal expenditures propensities. The use of the latter would increase the realism of the SAM approach. Thus, in the case of the consumption pattern of the socio-economic groups ($T_{2.3}$ in Table 6), the actual income elasticities of demand faced by the various households could be approximated by way of Engel's curves and corresponding marginal consumption propensities. The mapping of factorial income into household income ($T_{2.1}$) reflects the asset and wealth distribution, i.e. the ownership pattern of factors of production by household groups. As long as this pattern remains constant, the corresponding average propensities ($A_{2.1}$) appearing in Table 5 remain in effect. Conversely, should a change in the underlying pattern of ownership of factors occur—such as a land reform—this would have to be reflected by a new matrix of marginal propensities incorporating the structural change.

In a somewhat similar vein, certain types of technological changes—to the extent that they may be foreseen—could be incorporated into a matrix of marginal expenditures coefficients with regard to both the allocation of value added generated by production activities accruing to factors and the matrix of intermediate inputs.

The adoption of marginal expenditure coefficients would add realism to the estimates of the macroeconomic effects of alternative product-cum-technology activities within a SAM framework. There is, however, another restrictive assumption in the way the SAM was used in the above Korea case to explore the consequences of different activities on the policy objectives. Prices were assumed constant which, in turn, only holds true if excess capacity exists in the economy. At this junction of the state-of-the-arts in macroeconomic modelling, the building of a price-endogenous general equilibrium approach is an extremely difficult undertaking—particulary in view of the inadequacies of the theoretical hypotheses explaining investment determination, technological change and the operation of labour markets. The results obtained from such price-endogenous models might therefore be suspect to the extent that they are influenced (if not predetermined) by theoretically weak models embracing the above issues. Rather than attempting to build such models, the much simpler SAM approach, presented above, could be modified to incorporate sectoral capacity limitations and constraints on certain types of factors to render it more realistic. In a limited way an attempt might even be made in the next phase to explain some price changes—but certainly not all.

The final point, relating to future work, is the desirability of integrating the decision model developed in an earlier section at the microeconomic level with the macroeconomic SAM approach in this section. One interesting area of application which suggests itself is to postulate different macroeconomic preference functions held by different groups (e.g. the government, labour unions, employers, agricultural labourers) and explore the consequences of alternative output-cum-technology mixes (and policies affecting the latter) on these preference functions. We intend to address this issue in the next phase of our work.

APPENDIX

IN THIS APPENDIX we formally derive the proof of the *Proposition* presented earlier. The framework has the following elements:

1. There are ι goals, $G = (G_1, G_2, \ldots, G_\iota)$, where ι is a positive integer.

2. There are n decision-makers, $N = (N_1, N_2, \ldots, N_n)$ and each decision-maker advocates one or more of the ι goals. If more than one agent advocates a given goal, the weight assigned to that goal reflects the preference and relative bargaining power of all those who "push" the fulfilment of that goal.

3. Each goal G_i has a non-negative weight γ_i, the vector of weights being $\Gamma = (\gamma_1, \ldots, \gamma_\iota)$. Since only relative weights matter, they are normalized so that

$$\sum_{i=1}^{\iota} \gamma_i = 1.$$

If there is only one decision-maker the weights represent the decision-maker's preferences for the fulfilment of the various goals. If there are ι decision-makers and each one is exclusively interested in the fulfilment of one particular goal, γ_i may be thought of as the relative bargaining power of decision-maker (agent) i. In general, the weights are the product of the priority and relative bargaining power of the relevant decision-makers. For example, with a bargaining power of 1/2 and priorities 1/3 for G_1 and 2/3 for G_2, the decision-maker provides a weight of $1/6 = 1/2 \times 2/3$ for G_2.

Proof of the Proposition. The resources and the technological possibilities of the economy (production unit) determine the set S of all possible solutions (all possible degrees to which the various goals can be fulfilled). Mathematically, this set is convex and compact and lies in the non-negative orthant of R^{ι}_+. Since S is compact and convex, its boundary can be expressed in implicit form as a concave "goal fulfilment" frontier:

(A1) $S^* (G_1, \ldots ,G_{\iota}) \leqq 0, S^*_i > 0, i = 1, \ldots , \iota.$

In what follows we shall focus on the case when the frontier is concave but not strictly concave. The strictly concave case requires a more lengthy and tedious treatment, which can be found in Svejnar (1979 a). Axioms A1 and A2 guarantee that the solution U(S) (the observed fulfilment of G) is unique and lies on the frontier S^*. While the frontier S^* is concave,

the set $H = \left\{ G/ \displaystyle\prod_{i=1}^{\iota} G_i^{\gamma_i} \geq r \right\}$ is convex and its hyperbole boundary

$H^* = \left\{ G/ \displaystyle\prod_{i=1}^{\iota} G_i^{\gamma_i} = r \right\}$ is strictly convex in R^{ι}_+, for r ε R_+.

Let $(r^*) \subset R_+$ be the set such that $S \cap H \neq \phi$. Then $\dot{\Psi} (r^*) = \max (r^*)$ defines the unique point of tangency between H^* and S^*. It remains to be shown that $\Psi (r^*)$ and U(S) coincide.

At the unique point of tangency defined by $\Psi (r^*)$ the marginal rate of substitution (MRS) of G_i for G_j $(i, j = 1, \ldots , \iota; i \neq j.)$ along H^* equals the MRS of G_i and G_j along S^*. The latter MRS is given by the implicit function rule as

(A2) $\dfrac{\partial S^*/\partial G_i}{\partial S^*/\partial G_j} = \dfrac{S^*_i}{S^*_j} = - \dfrac{\partial G_j}{\partial G_i}, \ i,j = 1, \ldots , \iota; i \neq$

The MRS along H^* is given by

(A3) $- \dfrac{\partial G_j}{\partial G_i} = \dfrac{G_j}{G_i} \cdot \dfrac{\gamma_i}{\gamma_j}, i,j = 1, \ldots , \iota; i \neq .$

A comparison of equations (A2) and (A3) under the axioms of A3 (measurability) and A4 (proportionality) reveals that U (S) and $\Psi (r^*)$ coincide, as was to be shown.

Q.E.D.

Notes

[1] A framework for analyzing the various institutional factors influencing the state of technology in a few country-specific cases is included in the Final Report of the Cornell Project to the National Science Foundation. This report will also contain a part on the different views of "technology" and "techniques" espoused by different disciplines.

[2] For a recent survey of the available evidence on appropriate factor proportions in manufacturing see L. White (1978).

[3] For individual examples of these approaches see G. K. Boon (1975), H. Pack (1976) and G. Ranis (1977).

[4] See F. Stewart (1973 a, b) and P. Streeten (1972, 1973).

[5] The term acceptable is used here in the sense that none of the agents who could prevent the outcome or (impede) its implementation actually does so.

[6] This standardization is permissible since only relative weights matter. The advantage of course is that one can think of Γ_1 as the relative priority (weight) that G_1 has, in percentage terms, among all other goals. For example, if $\gamma_1 = 0.20$ then goal 1 has 20% of the total weight and goals 2 to ι get jointly 80%.

[7] The reader is referred to L. White (1976) for another example in the preference (goal) space.

[8] See Pyatt-Thorbecke (1976) for a basic explanation of the SAM approach and its use in development planning.

[9] The assistance of Sherman Robinson in providing the initial data set is gratefully acknowledged.

[10] The actual number was slightly less because some sectors do not include all four firm-sizes.

[11] Curiously enough two of these sectors display rising shares of capital value-added to total value-added as firm-size increases. This phenomenon is difficult to explain but might reflect the much greater efficiency of the technology adopted by large and medium size firms. It could also reflect differences in the product-mix of larger firms compared to smaller firms.

[12] Construction is an exception. One possible hypothesis is that smaller firms in the construction industry may tend to be engaged in carpentry and more skilled activities than large firms.

[13] T_{nx} embraces accounts 4, 5 and 6 (i.e. columns 61 and 63) and rows 1 to 59 and T_{xx} embraces the same columns and rows 61, 62 and 63 of Table 5.

[14] The actual transaction matrices T_{nn} and T_{kn} can consequently easily be obtained by multiplying the average propensities appearing in Table 5 by the corresponding column sum which is given in the total expenditures in row 65 of that table.

[15] The above multipliers appear in, respectively, rows 1, 2, and 3 of columns 36 in Table 8.

[16] Employment does not appear explicitly in the SAM Table 5 and 8. However, it is easy to derive the employment effects from the disaggregate output effects.

[17] In the case of the household consumption pattern, for example, this implies unitary elasticities of demand.

References

Adelman, I. and Robinson, S. (1978), *Income Distribution Policy in Developing Countries, A Case Study of Korea,* Stanford University Press and World Bank, 1978.

Bell, C. (1972), "The Acquisition of Agricultural Technology: Its Determinants and Effects", *Journal of Developing Studies,* October, pp. 123-59.

Boon, G.K. (1975), "Technological Choice in Metalworking, with Special Reference to Mexico", in Bhalla, A.S. (ed.) *Technology and Employment in Industry,* Geneva: International Labour Office.

Day, R. and I. Singh (1977), *Economic Development as an Adaptive Process,* Cambridge U. Press.

Feder, G. (1978), "Farm Size, Risk Aversion and the Adoption of New Technology Under Uncertainty", *Oxford Economic Papers,* Vol. 32, No. 2, July 1980.

Pack, H. (1976), "The Substitution of Labour for Capital in Kenyan Manufacturing", *Economic Journal,* Vol. 86, pp. 45-59.

Pyatt, G. and E. Thorbecke (1976), *Planning Techniques for a Better Future,* ILO, Geneva.

Ranis, G. (1977), "Industrial Technology Choice: A Review of Developing Country Evidence", *Interciencia,* February.

Rhee, Y.W. and L. Westphal (1977), "A Microeconometric Investigation of Choice of Technology", *J. Dev. Econ.* 4 (3), pp. 205-231.

Roth, A. (1977), "Individual Rationality and Nash's Solution to the Bargaining Problem", *Mathematics of Operations Research,* Vol. 2, No. 1, pp. 64-65.

Schweitzer, G.E. and Long, F.A. (1979, a), "Contributions to Industrial Development of Science and Technology Institutions in Malaysia, Nigeria and Colombia and Opportunities for Bilateral Cooperation (A Cross-Country Comparative Analysis)", Final Report to U.S. Dept. of State.

Schweitzer, G.E. and Long, F.A. (1979, b), "Developing a Methodology for Assessing the Contributions to Industrial Development of Science and Technology Institutions in Developing Countries", Final Report to U.S. Dept. of State.

Stewart, F. (1973, a), "Economic Development and Labour Use: A Comment", *World Development,* Vol. 1, pp. 25-28.

Stewart, F. (1973, b), "Trade and Technology", in Streeten, P. (ed.) *Trade Strategies for Development* New York: John Wiley & Sons.

Streeten, P. (1972), "The Multinational Corporation and the Nation State", in Streeten, P. (ed.) *The Frontiers of Development Studies* New York: John Wiley & Sons.

Streeten, P. (1973), "The Multinational Enterprise and the Theory of Development Policy", *World Development,* Vol. 1, pp. 1-14.

Svejnar, J. (1980a), "On the Empirical Testing of the Nash-Zeuthen Bargaining Solution", *Industrial and Labour Relations Review,* Vol. 33, No. 4.

Svejnar, J. (1980b), "The Bargaining Problem with Variable Bargaining Powers", Paper presented at the 1980 North American Meetings of the Econometric Society, Denver, September 1980.

Wells, L.T. (1975), "Economic Man and Engineering Man: Choice of Technology in a Low Wage Country", in Timmer *et al.* (eds.). *The Choice of Technology in Developing Countries.*

White, A.J. (1976), "Appropriate Technology, X-Inefficiency, and Competitive Environment: Some Evidence from Pakistan", *Quarterly Journal of Economics,* Vol. 90, No. 4.

White, A.J. (1978), "The Evidence on Appropriate Factor Proportions for Manufacturing in LDCs: A Survey", *Development and Cultural Change,* October, pp. 27-59.

Part II

Technological Development and Change

Editor's Introduction

ENVIRONMENTAL CONSTRAINTS on technological development are more ultimate in nature than the social, political and cultural constraints that are usually considered when adopting planning strategies for the developing world. It is very difficult to incorporate long-range ecological variables into the more traditional proximate framework. In order to develop realistic planning strategies for the next twenty years in the Third World, it is necessary to consider the long-term environmental consequences of change.

Man has been altering ecological systems on earth for thousands of years. As human population increased in many areas of the world, man has been manipulating the successional development of natural ecosystems in order to harvest as much food energy and nutrients as possible. Man, through his farming practices, including shifting cultivation and lowland delta cultivation, has tried to maximize net productivity. At the same time, attempts have been made to minimize the destruction of natural processes to ensure a reliable pulse of nutrients that could be harvested for a reasonable period of time.

In his modern agro-ecosystems, man still tries to maximize net productivity, i.e. the energy fixed at the first trophic level, whereas natural ecosystems operate within different parameters, maximizing for gross productivity, i.e total energy produced at the autotrophic level. Man selects crops that can harness as much solar energy as possible while using as little energy as possible for maintenance and respiration. Maintenance energy is the energy required to interact successfully with potential predators, competitors and pollinators. Because very little energy is generally made available for these maintenance activities, man has to "artificially" provide energy in the form of pesticides, herbicides and fertilizers to protect crop plants in this simplified biological community.

Environmental problems associated with agriculture are not new. With the advent of agriculture in the lowland delta regions of Mesopotamia for example, salinity caused crop losses that led to periodic food shortages. With the expansion of shifting cultivation through the upland areas of Africa and Asia, forest and grassland plant and animal extinction occurred periodically. The current intensification of agricultural practices throughout the world has led to an increase in the magnitude of environmental impacts.

In a recent United States Inter-governmental study on the state of global environmental quality, (U.S. Government Documents, 1980—The Global 2000 Report to the President, Volume II) it is reported that environmental, resource and population stresses are magnifying and will increasingly undermine the quality of human life on earth. The report states that the most serious environmental problem of the next two decades will be an acceleratory deterioration and loss of resources essential for agriculture, including the loss of crop land to erosion and deserts. The report concludes that by the year 2000, 40 per cent of the forests in the LDCs as of 1978, will be destroyed, world water supply per capita will decline by 35 per cent, and the atmospheric concentration of carbon dioxide will increase by about one-third its pre-industrial levels. Although the environmental problems projected for the last twenty years of the twentieth century are not unique, at no time in history has the biosphere been more vulnerable to ecological disruption than it is now.

With the alteration of relatively complex biological communities into monocultures to provide food and materials, man replaces inherently stable ecosystems with simplified systems that are vulnerable to pests, weeds, diseases, soil erosion and a variety of physical and climatic effects. Instead of recognizing that agro-ecosystems are subject to the same internal and external climatic, edaphic and biological variables that ultimately regulate function, man continues to maximize yields without regard for ecological determinants.

What is now needed is a readjustment in development strategy. Technological development programmes must explicitly take into account ecological variables that shape tropical agro-ecosystems and environmental limitations that impede agricultural intensification efforts. It may be necessary to trade-off some short-term economic gain for long-term environmental stability. Ecologically disruptive procedures such as the indiscriminate use of agricultural chemicals should be gradually phased out and sound environmental practices such as integrated pest management should be incorporated in their place. It will not be necessary to abandon modern capital-intensive techniques to improve yields, but what will be required is a gradual transition towards more intermediate technologies.

Basic to any readjustment is the need to recognize that agriculture is primarily ecological. An ecological strategy for agriculture would necessitate a detailed understanding of the dynamics of natural ecosystems. Internal homeostatic processes that have shaped traditional agro-ecosystems must be identified and maintained. Inherent mechanisms that promote biological productivity, ecological diversity and soil fertility, if retained, can serve to minimize environmental disruption. The capital-intensive technology strategy of the last two decades has led to improvements in per acre productivity. The transition to "ecologically sound" programmes will minimally reduce proximate production, but the short-term decline will be offset by long-term efficiency gains. The short-term difficulties that may result will be far less severe than the potential consequences of large-scale environmental deterioration.

Agricultural planning strategies that were adopted in the sixties did meet the goals set out for that decade. Food production has increased considerably over a relatively short period of time. Unfortunately yields cannot continue to increase unless modifications are carried out immediately.

Environmental variables have only recently been considered in the formulation of planning strategies. It has become clear over the past decade that environmental constraints, such as energy supplies and the availability of fresh water, could become the ultimate determinants of technological development throughout the Third World. If the ecological ramifications of planning programmes are given serious consideration, there is every reason to be optimistic about development efforts in the future.

Stephen Freedman

CHAPTER 4

Environment Implications of Development for the Third World

Margaret R. Biswas
Balliol College, Oxford University
and
Asit K. Biswas
President, International Society for Ecological Modelling

INTRODUCTION

DURING THE LAST DECADE the manifold development problems of developing countries have been the subject of numerous studies and discussions at various international fora. In spite of the serious attempts made by national governments, international organizations and bilateral aid agencies, overall developments in developing countries have left much to be desired and certainly have not approached the targets established by the First and Second Development Decades of the United Nations. Accordingly, Bradford Morse, Administrator of the UN Development Programme states in his latest Annual Report (1980) of the Agency:

> From one point of view, the 1970s was a decade of disappointments. Adequate gains were not made against poverty and its life-crushing consequences. The global economy fell short of the sustained expansion necessary for moving with much greater speed and effectiveness in the struggle to substantially ease hunger, disease, illiteracy, unemployment and lack of adequate housing. The world became joltingly aware that there were limits to its exploitable resources. Perhaps most frustrating of all was the fact that the industrialized and the developing countries did not achieve greater understanding—much less agreement—about how to deal with these problems effectively and equitably.

Few people will disagree with the above assessment. This is not to imply that no progress was made during the first two development decades, but rather that advances did not meet expectations. Accordingly, at the Third

117

General Conference of the UN Industrial Development Organization (UNIDO), held in New Delhi in early 1980, many governments expressed the opinion that the "two United Nations Development Decades had failed in their objectives" (UNIDO, 1980). Current trends indicate that the objectives of the 1975 Lima Declaration and Plan of Action that the developing countries attain a 25 per cent share in total world manufacturing output by the year 2000 will remain a dream. While developing countries have steadily increased their share in world manufacturing output since 1960, such gains have remained minor and are unlikely to generate major changes in regional development patterns. Preliminary estimates (UNIDO, 1981) indicate that in 1980 the developing countries' share of world manufacturing was 10.9 per cent compared with 8.2 per cent in 1960, 8.8 per cent in 1970 and 10.3 per cent in 1975. But the developing countries' share of the global population increased from 57.4 per cent in 1960 to almost two-thirds at present. At the present rate of development, the share of developing countries in world manufacturing output might not exceed 13 per cent by the end of the present century—a figure that is only about half the accepted target. The UNIDO Conference at New Delhi further noted that during the last two decades, "the rich became richer and the poor poorer; more than one quarter of the world's population was growing steadily poorer. Eight hundred million people, or about 40 per cent of the population of the developing countries, continued to live in absolute poverty; roughly a billion people lacked at least one of the basic necessities of food, water, shelter, education or health care" (UNIDO, 1980).

Nor can development in the agricultural sector be considered a success story. The target for the annual average growth rate of agricultural production for the Second Development Decade was established at 4 per cent, but the real average annual growth rate was only 2.8 per cent. The actual growth rate for many developing countries was much less. Furthermore, certain indicators of agricultural production in developing countries showed actual declines during the two Development Decades. For example, developing countries, in aggregate, were net exporters of grain in the 1950s. At the end of the first Development Decade, the surplus had become a net deficit. Developing countries as a whole imported 42 million tons of grain in 1970; this increased to 80 million tons by 1979. Current estimates of total grain import needs by the end of the third Development Decade in 1990 range from 125 to 150 million tons.

Similarly, if the index of per capita food production is considered, the situation is not much better for the low income developing countries (defined by the World Bank in 1980 as having per capita incomes of $360 and below in 1978). Of the 38 such countries listed in the World Bank's *World Development Report, 1980,* the index of per capita food production declined for 27 countries, remained the same for four, and increased for only seven countries. The index (1969-1971 = 100) declined to a low of 57 for Kampuchea, 71 for Mauritania and 80 for Togo and increased to a high of 114 for Sri Lanka, 108 for Sudan and 107 for Burundi in 1976-1978.

There are many reasons for such a dismal overall performance. Maurice J. Williams, Executive Director of the World Food Council, recently described the two decades of efforts to develop the agricultural sector as "twenty years of neglect" since developing countries did not give the sector the necessary priority in their national development plans.

ENVIRONMENT AND DEVELOPMENT DURING THE 1970s

FACED WITH SUCH DISMAL PROGRESS and several global crises, a series of world conferences was convened in the seventies by the United Nations and the highest political levels. The first of such world conferences was on the Human Environment, held at Stockholm in 1972. It was followed by conferences on Population (Bucharest, 1974), Food (Rome, 1974), Human Settlements (Vancouver, 1976), Water (Mar del Plata, 1977), Desertification (Nairobi, 1977), Agrarian Reform and Rural Development (Rome, 1979) and Science and Technology for Development (Vienna, 1979).

During the preparation of the Stockholm Conference on the Human Environment, it quickly became evident that environment and development problems were inseparable. Unfortunately during the late sixties and early seventies, our understanding of the multidimensional implications of development processes and their interrelationship with the environment was still somewhat rudimentary. Environmental protection was in the process of becoming a major national concern in certain countries. Before this period, overriding priority was placed on the first-order effects of technology and economic growth. Consequently, if there was a conflict between having more development projects at the cost of environmental degradation, it was almost axiomatic that it would be resolved in favour of the former. The secondary effects of development projects, e.g. increasing environmental pollution, would have been accepted as "the price of progress". It is still not unusual to find this philosophy among many professionals in many countries.

During the late sixties, the undesirable side-effects of economic development became highly visible and pronounced in many industrialized countries. Among the problems encountered were extensive air, water and noise pollution, difficulties with the disposal of solid wastes, lack or neglect of land-use planning, and a general deterioration of the quality of life in most urban areas. People became concerned with the problems of environmental deterioration and their resultant effects on health and lifestyle, and forcefully started to express those concerns. It was increasingly realized that human activities, based on the massive leverage which science and technology had made available, had reached a scale and intensity at which they were significantly modifying many of the elements of the biosphere that are vital in sustaining human life. Increased consumption of fossil fuels, proliferation of nuclear reactors, accelerated deforestation and loss of productive soil, intensification of water pollution, introduction of more than 1,000 man-made chemical compounds every year, loss of genetic diversity and other similar problems—all had impacts on the natural ecosystems in ways which even today cannot be fully evaluated. Aware of the fact that such development actions were creating real and, in many cases, immediate problems which often transcended national boundaries, the United Nations convened the World Conference on the Human Environment in Stockholm in 1972.

The discussions of environmental problems in developing countries of the late sixties and the early seventies were very much influenced by the experiences of the industrialized countries. The emphasis was primarily on the physical environment; there was very little analysis or understanding of the underlying socio-economic reasons for environmental deterioration. Two aspects generally received increasing attention in the West; constantly increasing use of raw materials and energy in developed countries and the "population explosion" in the Third World. Many doomsday scenarios were put forward which received extensive attention in the mass media.

On the question of depletion of resources, two reports were published in 1972, one by the journal *The Ecologist* entitled "Blueprint for Survival" and the other entitled *The Limits to Growth* by Meadows *et al.* Both these reports were based on the fact that exponential growth in a finite environment cannot continue indefinitely. *The Limits to Growth* made a profound impression on many people, and provided a bandwagon for parties who were convinced that mankind was headed for disaster unless growth-oriented policies were forsaken (Biswas, A.K., 1979a). Discussions of physical limits to growth became fashionable, without much discussion of social, institutional and technological factors that affect such growth and its limits.

Similar doomsday scenarios were put forward about the population sector of developing countries. Paul Ehrlich (1968) in his book *The Population Bomb* asserted that "the battle to feed all of humanity is over. In the 1970s the world will undergo famines—hundreds of millions of people are going to starve to death." Similarly William and Paul Paddock (1967) in their book *Famine 1975* advocated the policy of "triage—letting the least fit die in order to save more robust victims of hunger". They classified the three categories of hunger victims as "can't be saved", "walking wounded" and "should receive food". Countries like India and Egypt were classified in the "can't be saved" category. In a later edition of the book published in 1976, the Paddock brothers claimed that the "basic facts have not changed".

These types of statements, and clamour for "no-growth" in the West did not contribute to a mutual understanding of environment and resource problems between developed and developing countries. People in developing countries could not comprehend how their impoverished citizens could compete with the affluent consumers of the industrialized countries in using depleting global resources. In many cases they were not even aware of the resources available in their own countries since many regions were unexplored.

Accordingly, some developing countries felt that the concept of global resources management was an attempt to take away their control of national resources. Furthermore, since industrialized countries used the lion's share of resources and contributed to most of the resulting industrial pollution, developing countries did not see much reason to find and pay for solutions to these problems (UNEP, 1978).

Thus, in 1970, during the preparation of the Stockholm Conference, it became evident that many countries, both developed and developing, believed that environmental protection and development were not compatible, even though they subscribed to this view for very different reasons. Accordingly, in June 1971, the Conference Secretariat convened a seminar on "Environment and Development" at Founex, Switzerland. This seminar clarified several conceptual problems, the most important of which was that "environmental policies are integrated with development planning and regarded as a part of the overall framework of economic and social planning", and that "the developing countries should include environmental improvement as one of the multiple goals in a development plan and define its priority and dimensions in the light of their own cultural and social values and their own stage of economic development" (Founex, 1971). The Founex Report, the Stockholm Conference and the work of the United Nations Environment Programme, which was directly established as a result of the Stockholm Conference, clearly established that environment and development problems are compatible, and that the false dichotomy of "environment versus development" should no longer be recognized.

These events also focussed attention on the fact that many of the environmental problems of developing countries originated from under-development, i.e. lack of potable water and sanitation, squatter settlements, lack of education, employment and transportation. "Pollution of poverty" became an important theme at Stockholm. The concept that environment and development objectives are harmonious and mutually reinforcing received a further boost at the symposium on Resource Use, Environment and Development Strategies, jointly convened by UNEP and UNCTAD in 1974 in Cocyoc, Mexico, which resulted in the Cocyoc Declaration. This as well as other activities of the UN Environment Programme under such labels as ecodevelopment, environmental management and environment and development have significantly contributed to our understanding of population-resources-environment-development interrelationships, and it is now widely accepted that environment and development are two sides of the same coin.

A CASE STUDY: THE ASWAN DAM

THE ENVIRONMENT-DEVELOPMENT interrelationships can be best illustrated by a case study—The Aswan Dam in Egypt. The dam, built on the River Nile, is one of the largest in the world. It was built primarily for generating hydropower and was completed in 1968. It has received an extensive amount of criticism for contributing to environmental disruptions. A detailed analysis of the benefits and the costs of the Aswan Dam has yet to be made, but many of these effects can currently be perceived.

One effect is related to silt. Before the dam was constructed, large amounts of silt were either deposited on the Nile Valley or carried to the delta and the sea. These sediments are now being trapped in the reservoir created by the dam. Before the dam was built, suspended matter in the River Nile passing the Aswan ranged between 100 and 150 million tons per year. Observations made during the first few years after the completion of the dam indicate that the reservoir is losing about 60 million m^3 of storage per year due to siltation. At this rate, the dead storage capacity of 30 km^3 will be filled in about 500 years.

As a result of siltation in the reservoir, clear water is now flowing downstream to the dam causing erosion to the river bed and banks. One possible method to reduce erosion now being considered is to construct a number of barrages to reduce the velocity and force of the clear water. These barrages could also be utilized for power generation. The other possibility is to spill the water into Toshka Depression located to the west of the reservoir (Hafez and Shenouda, 1978).

Another effect of the siltation in the reservoir is the erosion of the Nile delta, some 1000 km away. Prior to the construction of the dam, the Delta used to be built up during the flood season with the silt carried by the Nile to the Mediterranean. This siltation in the Delta compensated for the erosion that resulted from the winter waves of the preceding year. Without enough siltation, erosion of the Delta has become a major problem, and studies are now being carried out to find a suitable solution.

Loss of silt has also affected the productive capacity of the Nile Valley which used to receive regular deposits of sediments each year. Currently, studies are being undertaken to assess the actual nutritive value of the silt, and the trace elements it contains, so that its loss can be compensated for by using chemical fertilizers.

Lack of sediments downstream from the dam has contributed to a significant reduction in plankton and organic carbons. This has, in turn, reduced the sardine, scombroid and crustacean population of the area. Loss of sardine along the Eastern Mediterranean has created economic problems for the fishermen who used to depend on the catch for their livelihood. Furthermore, there was a thriving small-scale industry making bricks from silt dredged from the canals. In the absence of such silt, many firms have now resorted to using the topsoil near the canals to make bricks, thus contributing further to the loss of productive soil in the country. Egyptian researchers have now succeeded in making bricks out of sand, but it will be some time before local industry can be persuaded to change from using topsoil. On the positive side, lack of silt has reduced the cost of dredging canals.

Besides siltation, other environmental problems created by the Aswan

Dam are the change of a terrestrial system to an aquatic system, hydrometeorological effects, and changes in soil and water quality. The High Dam created a vast reservoir with a shore-line length of 9250 km, surface area of 6216 km² and a volume of 156.9 km³ at 180 m elevation. It changed 500 km of the River Nile from a riverine to a lacustrine system. Though much of the land inundated was thinly populated, it contained areas rich in historical monuments, foremost of which was the Abu Simbel Temple. The temples of Abu Simbel and Philae (near Aswan) had to be dismantled and moved to higher locations. The huge man-made reservoir also changed the micro-climate of the area. It was calculated that the raising of the water level by 20 m, from 160 m to 180 m, more than doubled the lake surface, from 2950 km² to 6118 km², which increased the total annual evaporation from 6 km³ to 10 km³ (Biswas, A.K., 1979b).

The construction of the High Dam and Canal system for irrigation has tended to increase the water table in many parts of Egypt. Such developments and the tendency to over-irrigate are contributing to an increase in soil salinity

Figure 1   Change in Groundwater Level, West Nubariya, Egypt
            October 1969 to January 1974

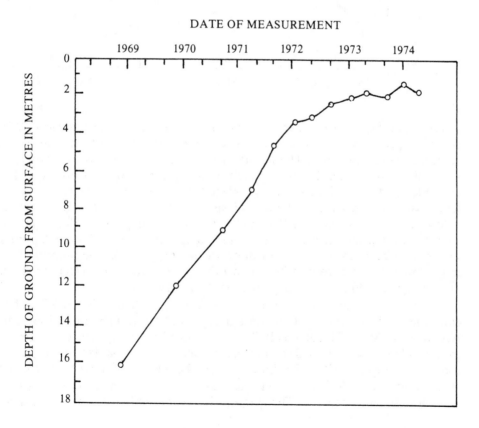

Figure 2   Increase in Salinity of Drainage and Irrigation Water
West Nubariya, Egypt, 1973 to 1975

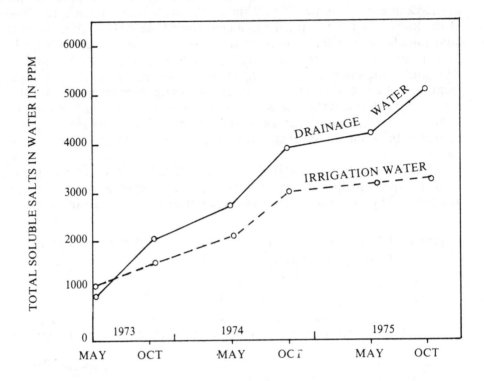

requiring an expensive and extensive construction of drainage systems. With
the disappearance of the annual Nile floods, the groundwater level has
stabilized at a higher level. The salinity in the irrigation canals is increasing and
some of the reclaimed lands are already facing a salinization problem. For
example, on the mechanized farm, West Nubariya sector, the groundwater
level rose from 16.2 metres below the surface in October 1969 to 1.3 metres
below the surface in January 1974, an average of .94 cm per day (Figure 1). The
salinity of the drainage water rose from 950 ppm in July 1973 to 5050 ppm in
October 1975, an average of 4.81 ppm per day (Figure 2). The salinity of the
irrigation water rose from 1150 ppm in July 1973 to 3200 ppm in October 1975,
an average of 2.40 ppm per day (Biswas, M.R., 1979a).

Loss of productive soil has become a major problem in Egypt which has
continued to lose some of its better agricultural land to urban development.
According to M.M. El Gabaly (1978), former Minister of Agriculture, Egypt is
losing 60,000 acres (120,000 cropped acres) per year to urbanization. In the 10
years from 1961 to 1970, 1.2 million acres, or 12 per cent of the total cultivable
land was lost. The 1972 law preventing use of agricultural land for urban uses is
now more strictly enforced. The magnitude of the problem can best be realized

by considering the fact that total irrigated land has virtually remained the same in Egypt during the last two decades, in spite of the thousands of hectares of new irrigated land developed due to the building of the Aswan Dam. Egypt has continued to lose good arable land to urbanization as fast as she has brought new land under irrigation, at tremendous investment costs. Overall, better quality agricultural land has been lost to urbanization than that brought under cultivation. Moreover, the agricultural land lost was closer to centres of population, and thus the energy cost of transporting products from the land to market and the necessity of building sophisticated storage systems was minimal. Since the new land is not as conveniently located, more energy has to be expended to transport, store and distribute the products, thus imposing additional cost on the economy (Biswas, M.R., 1979).

Replacement of simple primitive irrigation with perennial irrigation has led to a higher incidence of schistosomiasis. Where basin irrigation is still practised, the incidence is much lower. Data from the Egyptian Ministry of Health show that for the Asyut, Sohag and Qena Governorates, the overall prevalence rate for areas with perennial irrigation was 63.9 per cent (Asyut 68.1 per cent, Sohag 61.9 per cent and Qena 62 per cent) whereas it was only 16.2 per cent (Asyut 18.5 per cent, Sohag 10.4 per cent and Qena 13 per cent) in the basin irrigated areas (World Bank, 1976).

While much has been written on the negative side of the Aswan Dam, it should be recognized that its overall contribution to Egypt's economic develpment has been immense. It has provided irrigation for an additional 400,000 hectares of new agricultural land. Another 300,000 hectares of land in Upper Egypt, which was under basin irrigation and produced one crop every year, now produces two to three crops per year due to perennial irrigation. Productivity of existing land has improved (Abul-Ata, 1979). Navigation on the Nile and its tributaries, from the Aswan to the Mediterranean, has improved substantially. The project can now produce 10,000 million kwh/yr of electrical energy, equivalent to savings of 2 million tons of oil per year.

The total cost of the project was Egyptian £450,000,000, including subsidiary projects and extension of transmission lines. The total cost was recovered in only two years, at the rate of E £255,000,000 per year. Annual return to the national economy is as follows: E £140,000,000 from agricultural developments, E£100,000,000 from hydroelectric power generation, E £10,000,000 from flood control benefits and E £5,000,000 from navigation improvements. In terms of return on investments, the decision to develop the Aswan Dam was a good one. The real question is not whether the Aswan Dam should have been developed but rather what measures should have been taken to minimize the environmental costs to ensure that the benefits accruing from the project are not only maximized but also sustainable over the long-run.

CONCLUSION

THE EARLY MISCONCEPTION that environmental protection and development processes are incompatible has to be rectified. The concept of "environment versus development" needs to be replaced by a recognition of the complementarity between the environment and development. Development must be sustainable over the long-term; otherwise it will become a self-defeating process. Sustainable development (Biswas, A. K., 1980) can only be assured if development considerations are explicitly considered in the planning process.

References

Abul-Ata, A.A. (1979) "After the Aswan", *Mazingira,* Vol. 3, No. 3, pp. 21-26.

Biswas, Asit K. (1980) "Sustainable Development", *Mazingira,* Vol. 4, No. 1, pp. 4-13.

Biswas, Asit K. (1979a) "World Models, Resources and Environment", *Environmental Conservation ,* Vol. 6, No. 1, pp. 3.11.

Biswas, Asit K. (1979b) "Environment and Water Development in the Third World", *Journal of the Water Resources Planning and Management Division,* American Society of Civil Engineering, Vol. 106, No. WR1, pp. 319-332.

Biswas, Margaret R. (1979a) "Environment and Food Production", in *Food, Climate and Man,* Edited by Margaret R. Biswas and Asit K. Biswas, John Wiley & Sons, New York, pp. 125-158.

Biswas, Margaret R. (1979b) "Agriculture and Environment", Technical Memoir No. 3, International Commmission on Irrigation and Drainage, New Delhi, pp. 225-255.

Cocyoc Declaration, (1974) in "In Defence of the Earth", Executive Series No. 1, United Nations Environment Programme, Nairobi, 1981, pp. 107-119.

Ecologist (1972) "Blueprint for survival", *The Ecologist,* Vol. 2, pp. 1-43.

Ehrlich, P. (1968) "The Population Bomb", Ballantine Books, New York.

El-Gabaly, M. (1977) "Problems and Effects of Irrigation in the Near East Region", in *Arid Land Irrigation in Developing Countries,* Edited by E.B. Worthington, Pergamon Press, Oxford, pp. 239-249.

Founex Report on Environment and Development (1971) in "In Defence of the Earth", Executive Series No. 1, United Nations Environment Programme, Nairobi, 1981, pp. 3-38.

Hafez, M. & Shenouda, W.K. (1978) "The Environmental Impacts of the Aswan High Dam", in *Water Development and Management: Proceedings of the United Nations Water Conference,* Part 4, Pergamon Press, Oxford, pp. 1777-1785.

Meadows, D.H., Meadows, D.L., Randers, J. & Behrens, W.W. (1972) *"The Limits to Growth",* Universe Books, New York, 207 p.

Morse, B. (1980) "United Nations Development Programme in 1979", UNDP, New York 24 p.

Paddock, W. & Paddock, P. (1967) "Famine 1975!", Little Brown & Co., Boston, Republished in 1976 with a new Introduction and Postscript, 286 p.

Tolba, M.K. (1979) Foreword to "Food, Climate and Man", Edited by Margaret R. Biswas and Asit K. Biswas, John Wiley & Sons, New York, pp. xi-xvii.

UNIDO (1981) "A Statistical Review of the World Industrial Situation, 1980", Document UNIDO/ IS.214, United Nations Industrial Development Organization, Vienna.

UNIDO (1980) "Report of the Third General Conference, New Delhi, India", Document ID/CONF. 4/22, United Nations Industrial Development Organization, Vienna, 143 p.

World Bank (1980) "World Development Report, 1980", The World Bank, Washington, D.C., 166 p.

World Bank (1976) "Appraisal of Upper Egypt Drainage II Project, Arab Republic of Egypt", Report No. 1111-EGT, World Bank, Washington, D.C.

CHAPTER 5

Environmental Impacts of Agriculture:
Technical Aspects and Control Strategies

Pierre Crosson
Resources for the Future, Washington, D.C.

INTRODUCTION

THE EXPANSION of World agricultural production in the last several decades has resulted in rising environmental damage in many parts of the globe. Erosion from cropland and range reduces fertility and silts up reservoirs and irrigation systems; fertilizers and nutrients in animal wastes accelerate eutrophication of water bodies, endangering fish populations, and reducing recreational values, and nitrates in water may pose a threat to human and animal health; pesticides kill or injure unintended targets, including humans; the spread of irrigation promotes the spread also of schistosomiasis and other diseases, and creates problems of soil and water salinity; over-grazing in arid areas encourages desertification; and land clearing and drainage destroy plant and animal habitats, sometimes resulting in extermination of species with consequent diminution of the genetic pool on which all life depends (Eckholm, 1976).

For reasons given below we have no precise estimates of the social costs—measured by values lost—of the damage to the environment. The view is widely held, however, that the costs are sufficiently high to pose a threat to the long-run viability of the natural resource base in some parts of the world. Population growth and the rising demand for food, particularly in the developing countries, indicate that whatever the present environmental costs of world agricultural production, they are likely to increase significantly over the next several decades.

This chapter deals with the principal issues involved in controlling the environmental costs of agriculture within socially acceptable limits. The discussion is in three parts. The first deals with the nature of environmental costs. The second part considers measures for controlling environmental costs,

129

distinguishing between technologies of control and strategies to promote adoption of these technologies. The third part draws on the second to develop ideas for an alternative control strategy.

THE NATURE OF ENVIRONMENTAL COSTS

ALL PRODUCTION costs something, measured by the value of the resources used. Consequently, the environmental costs of agricultural production cannot *per se* be a matter of special social concern any more than the costs of fertilizer, labour, machinery services, and so on. Environmental costs *are* of special concern, however, because, unlike the costs of other resources, environmental costs are not properly taken into account in the production decisions of farmers and other managers of resources. The result is a pattern of resource allocation and of production which is socially inefficient and which may also be inequitable. Social inefficiency results because the failure to give proper weight to environmental costs means that at the margin the social costs of production exceed the social value of production. The equity issue may arise because environmental costs are not borne by those who generate them but by others, and in circumstances in which those bearing the costs find it difficult or impossible to exact compensation.

The difficulty in exacting compensation occurs because those bearing environmental costs have no enforceable property right in the land, water, air, or ecosystems which transmit or absorb the impact of the damaging materials generated by the farmer's operations. Consequently the bearers of the costs cannot deny the farmer access to these resources nor can they charge him for his use of or damage to the resources. This absence of property rights is at the core of both the efficiency problem and the equity problem represented by environmental costs.

It is important to note that the inefficient use of resources represented by environmental costs is only from the standpoint of society. From the farmer's standpoint the pattern of resource use can be assumed to be efficient. In his calculations he quite understandably ignores environmental costs because he does not pay them. He thus meets the conditions for an efficient allocation of his resources—that at the margin the cost of each resource equals its value in production—by treating environmental costs as zero.

The problem of environmental costs, therefore, reflects a divergence between the public and the private interest in the management of resources. In the economic literature on environmental costs, the focus typically is on the efficiency issue raised by this divergence of interests. That the divergence may also pose an equity issue is less commonly addressed. This perhaps is because equity is in the eye of the beholder. A farmer who always has irrigated with water made saline by the return flow of upstream irrigators may perceive no equity issue even though his yields are adversely affected by the salinity. If the situation is stable and of long standing, and the downstream farmer sees no way of altering it, he may regard it as part of the natural order of things. If the salinity is of recent origin, however, and the farmer observes a drop in his yields he likely will conclude that his upstream neighbours are treating him unfairly and may seek redress.

In agriculture the social problem represented by environmental costs is to somehow change the practices of the farmer, or other managers of agriculture resources, so that at the margin the cost of production, including environmental costs, is equal to the value of production; and to do this in a way that minimizes perceived inequities in the solution. Some approaches to this problem are considered in the next section.

CONTROL OF ENVIRONMENTAL COSTS

IT IS USEFUL to divide the control problem in two parts: development of control technologies, and adoption of the technologies by farmers, or by others where off-farm control are required. Knowledge of control technologies generally is much more advanced than is knowledge of how to promote adoption of the technologies.

Control Technologies

Utilization of known techniques and management practices will substantially reduce erosion under most circumstances, or prevent eroded soils from reaching water bodies. Terracing, contour ploughing, strip cropping, grassed waterways and conservation tillage, singly or in some combination, will reduce erosion 50-90 per cent, depending on specific conditions of soil, terrain and climate. Settling ponds and check dams, if properly constructed and maintained, are effective means of preventing movement of eroded soils to rivers, lakes or reservoirs.

Salt build-up on irrigated lands usually can be adequately controlled by proper water management. Leaching to flush salts out of the soil, combined with surface or sub-surface drainage generally are effective control measures. Of course the drainage waters will contain a relatively high concentration of salt, and the concentration will rise if the water is re-used for irrigation downstream. Thus effective control of salt build-up by any one farmer may simply transfer the problem to downstream users of the water. The drainage waters may eventually become so saline that they cannot safely be returned to the river. Some other means must then be found for their disposal, e.g. desalinization, as the United States is proposing to do with Colorado River water before it flows to Mexico, or dumping the water in the ocean or some other body of water large enough to receive it without damage.

Substitution of more salt resistant crops, e.g. barley for wheat, or breeding of more salt-resistant varieties of given crops also are measures for dealing with the salinity problem. These technologies, however, are less widely employed as controls than leaching and drainage.

In many of the developing countries the spread of irrigation has contributed to a rise in the incidence of schistosomiasis and malaria. In addition irrigation projects involving large reservoirs have destroyed riverine ecosystems, altered the timing and nutrient content of river flows to valuable coastal estuaries and lagoons, thus reducing their productivity, and have displaced large numbers of people. There are known techniques for controlling schistosomiasis and malaria—they involve destruction of the vectors transmitting the diseases and reducing exposure of people to them—but technical control of the other environmental costs of large reservoir based systems is more difficult and often impossible. A decision to build such systems usually is a decision to accept these costs.

The environmental damages from use of pesticides can be greatly reduced by proper care in application, by selective application, and by substitution of less damaging for more damaging materials. Simple procedures such as following application instructions on the container and wearing proper protective clothing will greatly reduce injuries suffered in application. Use of field scouts combined with knowledge of the life cycles of insects and plants permits more effective insect control with a given amount of insecticide or the use of less insecticide for a given amount of control, than conventional pest management practices. This technique has proved highly effective in controlling cotton insect pest in Texas. The substitution of synthetic pyrethroids, with relatively low toxicity to mammals, for highly toxic organophosphorus compounds also shows promise in cotton insect control. Since more insecticides are applied to cotton by far than to any other crop, the success of scouting and the apparent potential of pyrethroids in cotton are promising.

These practices for control of cotton insects are examples of Integrated Pest Management. IPM is a fuzzy concept. It embraces a variety of practices, but the core idea is that chemicals are employed selectively with scouting and perhaps more pest resistant crop varieties to hold pest damage at an economic threshold determined by crop prices and control costs. The emphasis in IPM is on flexibility in selection of control techniques. In effect it achieves pest control by substituting knowledge for rule-of-thumb application of pesticides. So far IPM has had its greater success in cotton and some fruits. Its potential with respect to other main crops such as grains and soyabeans, is uncertain.

The environmental costs of fertilizer can be significantly reduced by more careful and timely application and by erosion control. More exact placement of nutrients near the root zone at a time when crop demand for nutrients is greatest can increase the proportion of fertilizer applied which is taken up by the crop, thus meeting crop needs for nutrients while reducing the amount of these nutrients which escape to the environment. Erosion control also reduces the escape of nutrients to the environment, particularly of phosphorus since phosphorus has low water solubility and moves off the farm primarily by adsorption to soil particles. Erosion control is less successful in reducing escape of nitrates since these are water soluble and move to streams with run-off or to groundwater through percolation. Of course escape of nitrates can be reduced by practices which increase the proportion of given amounts of nitrogen applied which is taken up by the plant.

Technical measures for the control of animal wastes from feedlots are well known. The key is collection and impoundment of the wastes to prevent their transport to water bodies. Natural processes of course continually degrade the wastes, although this may create unpleasant odours—a form of air pollution. Care must also be taken to manage the impoundment to prevent percolation of unacceptable amounts of nitrates to groundwater.

Desertification can be halted, or even reversed by irrigation or development of ground cover better suited to the arid environment. The loss of plant and animal habitat because of land clearing and drainage, however, cannot generally be controlled by known technologies. The environmental costs of these activities, like some of those imposed by construction of large reservoir based irrigation systems, cannot be avoided, given the present state of knowledge.

In summary, with the exceptions just noted, we know a great deal about technical means for controlling the environmental costs of agriculture. Of course there are gaps in our knowledge, but there is no question that widespread application of known control technologies would greatly reduce environmental costs of agriculture all around the world. The key problem in controlling environmental costs, therefore, is not lack of technical knowledge of what to do. The problem is to achieve widespread application of the knowledge we have. Since the environmental costs of agriculture generally are perceived to be high and threaten to rise further, why is this so difficult to achieve?

Implementation of Control Techniques

Implementation of known technologies for control of environmental costs of agriculture is difficult because adoption of these technologies is not in the perceived interest of farmers and other managers of agricultural resources. Deployment of these technologies imposes costs on the farmer but since he does not pay environmental costs he receives no benefit from reducing them. This is a reflection of the divergence between the public and private interest in control of environmental costs. The key to adoption of control technologies is to create conditions causing these interests to converge.

There is a line of argument in the literature that the necessary convergence of interests can be achieved through private bargaining between those who impose environmental costs and those who bear them (Coase, 1960). The necessary condition for a bargain is that the marginal environmental costs exceed the marginal cost of reducing them. Given this condition the bearer of the costs has incentive to bribe he who imposes them to change his practices to reduce the costs. The bribe will be the amount necessary to compensate the imposer of the costs for his expenses in changing his practices so as to reduce marginal environmental costs to equality with the marginal cost of achieving reduction. This equality indicates that the bargaining process yields a socially efficient level of production and pattern of resource use.

Discussion of this argument subsequent to its initial presentation by Coase demonstrated that the bargaining process likely would founder or not even get under way if the numbers of those both imposing and bearing environmental costs are large and widely dispersed. Under these conditions, the costs of organizing the bargain would be high. Moreover, any particular individual among those bearing environmental costs would have incentive to hold off from organizational effort since he would share the benefits even if others did the work. For example, all downstream farmers benefit from reduced salinity in the river even if some of them do not participate in the organizational effort necessary to bring it about.

In addition the argument for private bargaining as the way to deal with environmental costs ignores the equity issue implicitly raised by the argument. A downstream farmer, for example, may know that he would be better off if he bribed upstream farmers to change their irrigation practices to reduce the salinity of the water he receives. However, if the river is common-property the downstream farmer may well view the imposition of damage on him by his neighbours upstream as arbitrary and unfair; and the suggestion that he can reduce these damages by bribing those upstream to stop doing something which in his eyes they have no right to do in the first place may strike him as outrageous. That the pattern of resource use and level of production resulting from a bargain would be socially efficient will probably appear irrelevant to the downstream farmer.

Apart from the equity issue the conditions under which environmental costs of agriculture occur in general do not favour solution through private bargaining. Both those imposing and those bearing the costs typically are numerous and geographically widespread. And of course there is no way future generations can participate in the bargaining process even though their interests may be adversely affected by permanent damage to the agricultural resource base.

If private bargaining cannot achieve congruence between public and private interests in control of environmental costs public intervention becomes necessary. And in fact public intervention is the mode of control adopted in the United States and other countries around the world.

The literature on public intervention to control environmental costs focusses almost exclusively on two control strategies: (1) regulation of practices; (2) financial incentives to change practices. The actions of the US Environmental Protection Agency with respect to pesticides illustrate the

by stringently regulating access to it. In these cases, implementation of the new technologies strategy may take too long to be effective. As this illustrates, commitment to the new technologies strategy would not imply abandonment of the regulatory or incentives strategies. There still would be instances in which those approaches would be more appropriate than the new technologies strategy.

So the new technologies strategy is not a panacea. It has significant advantages over the regulatory and incentives strategies, however. Its great strength is that it relies not on coercion or bribery but on the play of the farmer's own interest to achieve convergence between the social and private interest in the use of resources. Further, to the extent that it succeeds, the new technologies strategy diminishes the equity issue represented by the existence of environmental costs.

Is it reasonable to believe that as a society we can develop technologies which simultaneously satisfy the farmer's interest and society's interest in the use of agricultural resources? The answer surely is yes. We already have done it. Integrated management of cotton insect pests in Texas has sharply reduced the quantity of insecticides applied with no sacrifice in yield and no apparent offsetting environmental costs. The control programme was developed because the emergence of resistance to DDT and other insecticides by the boll weevil and tobacco budworm threatened the economic viability of chemically based control measures and because of concern about the environmental impacts of these measures. The control programme was adopted because farmers found it in their economic interest to do so. Neither coercion nor bribes were used.

Conservation tillage also appears to be a technology consistent with the new technologies strategy. Conservation tillage, by leaving much crop stubble on the soil surface, dramatically reduces erosion compared to conventional tillage, which ploughs the stubble under. This favourable environmental effect may be offset to some extent because conservation tillage relies more on herbicides and less on cultivation to control weeds. In general, however, herbicides impose relatively low environmental costs. Consequently, where erosion is high, conservation tillage looks promising. The technology has spread rapidly in the midwestern and southeastern United States since the mid-1960s, wholly because of its economic attractiveness to farmers.

Integrated pest management and conservation tillage demonstrate that technologies meeting the conditions for the new technologies strategy are possible. Can we expect such technologies to emerge without public intervention or encouragement? The development and spread of conservation tillage in the United States resulted largely from private initiative, indicating that public intervention is not a necessary condition for development of technologies consistent with the new technologies strategy. In general, however, public intervention would most likely be required. Private investment in development of new agricultural technologies is guided primarily by perceptions of what will be economically attractive to the farmer. Since environmental costs typically do not enter the economic calculations of the farmer they will not be properly reflected either in private decisions to develop new technologies. Pursuit of a new technologies strategy, therefore, will require a measure of public intervention.

Efficient allocation of public resources towards development of alternative new technologies requires criteria by which to judge the alternatives. One set of criteria is given by the condition that all alternatives must be at least as economically attractive to the farmer as the technology he currently uses. Other criteria are determined by the severity of the environmental problems requiring attention. In setting these criteria the new technologies strategy requires the same sorts of information about environmental costs as the regulatory and incentives strategies; and faces the same problems posed by the absence of a common denominator with which to measure these costs. In setting the environmental criteria, therefore, those developing a new technologies strategy would have no unambiguous standard by which to judge the relative importance of various environmental costs. In this respect the new technologies strategy shares the same weakness as the regulatory and incentives strategies.

Of course the new technologies strategy will not work unless public agencies responsible for allocating resources to research on agricultural technologies are sensitive to environmental problems and committed to the new technologies strategy as a mode for dealing with them. If these conditions are met, then there is every reason to believe that public investment in development of agricultural technologies will be directed to deal with emerging environmental problems of agriculture. The argument here is an extension of the Hayami-Ruttan argument that public investment in agricultural technologies is guided by emerging resource scarcities as signalled by rising prices of resources (Hayami and Ruttan, 1971). While researchers in public agencies and politicians have no immediate pecuniary stake in solving problems of increasing resource scarcity, they are induced to do so by the evidence of increasing scarcity and by the demands of farmers for technologies appropriate to the emerging scarcities. Hayami and Ruttan build their argument on careful analysis of American and Japanese experience in developing agricultural technologies.

The role of prices as signals of resource scarcity and the pressure of farmers for appropriate technological responses are crucial to the Hayami-Ruttan induced innovation argument. Since environmental costs are not priced, this element will not come into play to support development of a new technologies strategy. Moreover, so long as farmers do not pay environmental costs, they will exert little if any pressure for technologies to deal with them.

Despite these limitations, public agencies can be induced to develop technologies that reduce damages to the environment. While there are no prices signalling the emergence of environmental problems, there is much physical evidence: tons of top soil lost, accelerated siltation of reservoirs, increasing insect resistance to insecticides, rising nitrate levels in groundwater, and so on. Although assigning priorities among these problems is difficult, judgements can be and are made that some of them are serious enough to warrant public action. The pressure to take action may not come from farmers, but the strength of the environmental movement in the United States and other countries around the world shows that the pressure is there. Even farmers may bring pressure to develop new technologies if they see these as alternatives to burdensome regulatory programmes.

I have stated that the great strength of the new technologies strategy is that unlike the regulatory approach it works with rather than against the interests of the farmer. I suspect that the costs of the new technologies strategy also would compare favourably with the costs of the regulatory and incentives approaches, although I have no hard evidence on this score. To illustrate the point, however, I hypothesize that in the United States the investment of a few hundred million dollars to develop conservation tillage technologies adapted to specific resource conditions in areas where erosion now is or might become serious would have a far greater impact in reducing erosion than the same amount of money spent on regulatory or incentives programmes. I suspect also that the costs of a new technologies strategy to reduce environmental damage of pesticides would compare favourably with the public and private costs of the programmes now employed for that purpose. It is concluded that

In summary, the conditions for a new technologies strategy to control some important environmental costs of agriculture can be met. There is strong and reason to believe that such a strategy would be more effective than the regulatory strategy since it does not rely on coercion of farmers. Moreover, the new technologies strategy would reduce, if not avoid entirely, the equity problems not now adequately addressed by either the regulatory or incentives strategies. Finally, although I do not claim to have demonstrated this, the costs of the new technologies strategy likely would probably compare favourably with those of the regulatory and incentives strategies. Given the present prospective importance of the environmental problems of agriculture and the weaknesses of the regulatory and incentives strategies for dealing with them, the new technologies strategy merits close attention as an additional mode of control.

References

Coase, R.H (1960). The Problem of Social Cost. Journal of Law and Economics, No. 3, pp. 1-44.

Eckholm, E. (1976). *Losing Ground: Environmental Stress and World Food Prospects.* W.W. Norton.

Hayami, Y. and Ruttan, V. (1971). *Agricultural Development: An International Perspective.* Johns Hopkins Press.

Kneese, A.V. and Bowler, B.T. (1968). *Managing Water Quality: Economics, Technology, Institutions.* The Johns Hopkins Press for Resources for the Future. pp.101-109.

Sharp, B.M.H. and Bromley, D.W. (1979). Agricultural Pollution: The Economics of Co-ordination. *American Journal of Agricultural Economics,* Vol. 61. No. 4 pp. 591-600.

Key to Figures 1, 2 and 3

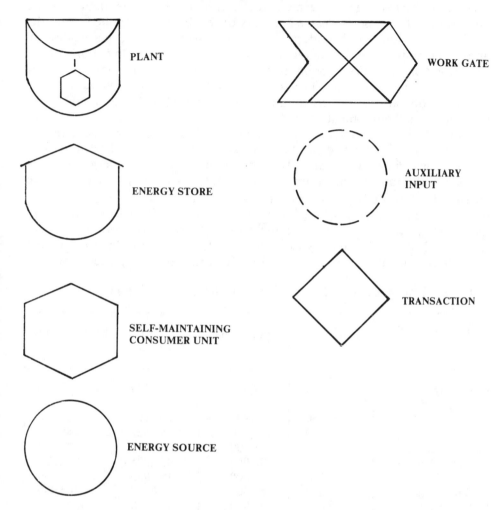

PLANT

WORK GATE

ENERGY STORE

AUXILIARY INPUT

SELF-MAINTAINING CONSUMER UNIT

TRANSACTION

ENERGY SOURCE

The third category is the "capital-intensive" method of rice production (see Figure 3) usually referred to as the "Green Revolution" style of agriculture. The modified operations centre around the adoption of high yield hybrid varieties, such as IR-8, that are dependent on substantial inputs of fertilizers, insecticides, herbicides and large-scale irrigation to produce the high yields. The land preparation, harvesting and threshing operations are improved and, where possible, tractors and threshers are used to perform the operations that were once done manually.

The final category is the "experimental" method of rice production (see Figure 3). This agricultural method as practised at various international agricultural experimental stations is still in the research stage and has not yet had widespread application. Improved high yield varieties of rice are used together with a package of intermediate chemical inputs. Rather than applying large doses of chemicals, lesser amounts of fertilizers are applied using the mudball technique, and moderate doses of insecticides and herbicides are applied in conjunction with integrated control of predators and competitors. The irrigation operations are improved with respect to timing and manner of application onto the field. The seeding, land preparation, harvesting and threshing operations are mechanized, but instead of large tractors and machines, small equipment is used, requiring less fuel and maintenance.

In a previous chapter, energy inputs and yield averages were calculated for each operation in the rice production system, for each method described above. (See Pimentel *et al.* 1974, for process method energy analysis procedures.) Human and animal labour, equipment and fuel costs were measured in terms of kilocalories per hectare, based on average requirements reported in the literature (Freedman 1978 and 1979). The calculated energy requirements and yield totals for the four methods of rice production are shown in Table 1. The energy input for traditional rice production totalled 327,910 kilocalories per hectare, the lowest of the four categories analyzed.

Table 1　Energetic Totals for Rice Production (Kcal/ha)

Process	Traditional	Transitional	Green Revolution	Experimental
Seeding	27,000	27,000	31,770	5,580
Land Preparation	109,650	373,900	496,340	283,650
Fertilization	9,360	111,390	1,954,627	1,146,467
Insect Control	-	108,000	65,365	45,117
Weed Control	-	72,000	435,395	156,027
Irrigation	3,600	36,000	1,183,980	609,990
Harvesting	131,500	180,300	167,847	104,990
Threshing	46,800	135,900	139,430	77,840
Other (Cost of Energy, Seed[1], Drying[2], and Transportation)	-	419,740	1,020,000	1,020,000
Totals:				
Input Totals	327,910	1,464,230	5,494,754	3,449,661
Yield Totals	4,537,500	9,720,000	15,972,000	15,972,000

1.　Cost of seed production based on estimates calculated by Pimentel *et al.* 1974.

2.　Energy estimates of drying and transportation based on an assessment done by the Food and Agriculture Organization reported in 1976.

The transitional method amounted to 1,464,230 kilocalories per hectare, more than three times the energy input of the traditional method. Differences in energy input between the two methods are most pronounced for the land preparation and auxiliary input operation.

The Green Revolution method requires the highest input, 5,494,754 kilocalories per hectare. Most energy is required for the fertilizer, herbicide and irrigation operations. The experimental method requires 3,449,661 kilocalories per hectare. Compared to the Green Revolution method, there is a substantial decrease in energy input particularly at the preparation and weeding stages.

Rice production methods can be evaluated in terms of energy efficiency by comparing energy utilization and the output of food produced. The less energy input, for a given quantity of food energy produced, the higher the efficiency of the particular method. Several broad generalizations can be derived from efficiency calculations reported in the literature. Traditional methods of agricultural production usually are the most efficient, offering the highest kilocalories return for a given investment of cultural energy (Black 1971 and Rappaport 1971). Although the traditional methods are the most efficient, most methods analysed demonstrated a net gain in kilocalorie production for each kilocalorie of cultural energy input (Matsubayashi 1963). Advanced methods of rice production previously analyzed (Freedman 1979), showed a relatively high return of kilocalories per unit of energy invested. The transitional method yielded 6.6 kilocalories for each kilocalorie input, and the efficiency of the experimental method was 4.63, far less than the transitional method, but still providing a net gain. The Green Revolution method was found to have the lowest efficiency of 2.9 which still can be seen as a moderate overall gain of energy per kilocalorie invested.

The investment of cultural energy (man's energy input) in the transitional, Green Revolution and experimental methods led to an increase in yields. The average yield for the traditional method is only 4,537,500 kilocalories per hectare. By increasing the cultural energy investment, the transitional method produced twice the yield while the yields for the Green Revolution and experimental method were more than three-and-a-half times those of the traditional method (see Table 1).

Obviously the increase in yields was largely the result of the increase in cultural energy from 327,910 kilocalories invested in the traditional method to 5,494,754 kilocalories invested in the experimental and Green Revolution methods. This seventeen-fold increase in cultural energy has led to an increase in food energy per hectare. It has been argued that presently this is the only option available to Third World agricultural planners. In the past, to increase the amount of food produced, more land was put under cultivation. Presently, very little additional arable land is available for use. Therefore, the increase in cultural energy can be seen as a strategy of "short-circuiting" the "more-people-more-agricultural-land" dilemma.

The Green Revolution method, although not as energy efficient, can be evaluated as optimal, if the goal is to increase yield on a per hectare basis. Even though the Green Revolution method is least efficient in terms of energy produced per unit of energy invested, it is the most productive in terms of energy produced per unit of land used. This is precisely the reason why capital-intensive technology continues to be developed in Third World agriculture.

COMPARATIVE OPERATIONAL EFFICIENCIES

ONE COULD ALTERNATIVELY proceed from the assumption that what should be measured is the way energy is actually used for the production of food. In other words, what matters is the efficiency with which cultural energy is invested, rather than the way energy is converted into food. Commoner (1973) suggested that it would be more informative to measure energy efficiency by the ratio of the actual amount of energy used to produce a given output, to the theoretical minimum amount of energy the task requires. Efficiency is then a measure of the way production is carried out relative to the optimal way the task can be performed.

Comparative efficiencies have been calculated for various agricultural methods (Freedman, 1980). The comparative efficiency for the transitional method is 47.97 per cent, roughly half that of the traditional method, suggesting that about 52 per cent of the potential energy input is not directly translated into food energy (see Table 2). The capital-intensive Green Revolution method was calculated as 21 per cent, thus about 80 per cent of the potential energy input is lost in food production process. For the experimental method, approximately two-thirds of the production energy is not transferred directly into food energy (see Table 2).

Comparative efficiency can also be calculated at each operational level, taking into account equivalent yield potentials (Freedman 1979). Table 3 shows the results of the comparative efficiency calculations for each of the rice production operations.

Using the operational comparative efficiencies, it is possible to point out areas for technological development. Although overall comparative efficiency was highest for the transitional method (47.97 per cent) it is possible to modify that method in such a way as to improve yields without the loss of overall efficiency. It is possible to maintain yields at levels equivalent to those of the Green Revolution method, by reducing energy use at certain levels of operation, such as irrigation, fertilization and weed control, by the transfer of labour previously required for land preparation, harvesting and threshing. These primary operations can be accomplished more efficiently by adopting the experimental style of operations at these previously "labour-intensive" levels of production.

The comparative efficiency calculations provide an information base for

Table 2 Overall Energy Efficiencies

	Average Output / Average Input	Ratio Minimum Energy / Actual Energy Used	Comparative Efficiency
Traditional	13.830	$\dfrac{327{,}910 \times 3.52}{327{,}910 \times 3.52}$	100.00
Transitional	6.638	$\dfrac{327{,}910 \times 3.52}{1{,}464{,}230 \times 1.64}$	47.97
Green Revolution	2.907	$\dfrac{327{,}910 \times 3.52}{5{,}494{,}754 \times 1}$	21.00
Experimental	4.630	$\dfrac{327{,}910 \times 3.52}{3{,}449{,}661 \times 1}$	33.64

Table 3 Comparative Efficiencies at Each Operational Level

	Seeding	Land Preparation	Irrigation	Fertilization	Pest Control	Weed Control	Harvesting	Threshing
Traditional	5.87	73.49	100.00	100.00	-	-	22.68	47.25
Transitional	12.60	46.26	21.42	18.00	25.47	100.00	35.51	34.93
Green Revolution	17.56	57.15	1.07	1.69	69.02	27.12	62.55	55.83
Experimental	100.00	100.00	2.08	2.87	100.00	75.68	100.00	100.00

the evaluation of energy efficiency, providing a framework for agricultural planners to set up technological innovations. Alternative procedures for each task should be evaluated in terms of comparative efficiencies, and the most effective combinations of labour and machinery can be decided upon after analyses of regional resources and labour availability. Labour input should be redirected to auxiliary operations where possible and technological modifications be primarily focussed on primary operations, where labour is now predominantly used.

SPECIFIC OPERATIONAL MODIFICATIONS

BASED ON THE efficiency calculations (Table 3) it is possible to recommend specific planning strategies for each operation. For example, at the seeding level it is clear that modern seeding drills which allow for the direct seeding of rice are the most appropriate modification of the seed broadcasting operation. This experimental technique will not displace labour, but will allow for technological improvement and increased energy efficiency. Currently there are many varieties of tube seeders available for use in the tropics, some on wheels, some with metering devices. The International Rice Research Institute has been working on tube seeder basic design for quite some time, and has produced an implement that works well in both wet and dry fields, spreading either sprouted or dry seeds. For the transplanting operations, transplanting forks and gravity type transplanters are being refined for flooded field conditions.

As far as land preparation is concerned, the experimental modifications clearly represent the most appropriate strategy for development. Small manually operated tillers are proving to be efficient land preparation devices (Moowaw and Curfs, 1972). "No tillage" or "conservation tillage" land preparation styles should continue to be studied for further adoption in the tropics (Triplett and Doren, 1977).

Two further primary operations, harvesting and threshing, should continue to be modified along the same experimental lines. Small self-propelled single purpose reaper-binders operated by 25 horsepower engines for example, seem to offer a viable alternative to the labour-intensive reaping presently employed. Mechanization has proved difficult because of field conditions at the time of harvest and because of generally small plot sizes. The wet field conditions provide for poor mobility and handling of reapers. Threshing by contrast has become increasingly mechanized in the Third World, particularly since the mid-sixties. Paddle threshers, operated manually, and various simple machine-operated, five to thirty horsepower threshers, are manufactured for Third World use. Farmers there can especially use lightweight, easy to operate and maintain power threshers. In particular, there is a specific need for a lightweight thresher, that can be easily moved around a field.

Modifications of "auxiliary operations" on the other hand should be more labour-intensive. For example, since labour would now be more generally available, intercultivation techniques could be employed for weed control allowing for the diminishing use of herbicides. Hand hoeing, machine discing and the use of mouldboard ploughs throughout the growing season can eliminate the need for the expansion of herbicide use. Biological control of weeds has long been neglected as an option which might prove as efficient as the integrative control of insects.

In the last few years there have been great strides taken in the development of integrated pest management programmes in the tropics. However, dusting still remains the preferred method of insecticide application in the developing world. If chemicals continue to be used in increasing amounts, greater effort will be required to develop more effective treatment techniques. Research is needed on paddy water and root zone application. However, there is still very little known about the actual mechanics of getting insecticide into the root zone, and equipment and technical knowledge are still lacking. With the development of new equipment, such as the water-band and the granular band applicators at IRRI, there will be a greater emphasis on direct application of insecticide. With the development of such newer devices as liquid root zone injectors, the time of application should be cut down considerably (IRRI, 1976) and direct application may become more widely accepted.

Labour-intensive operations should also be developed at the fertilizing stage of rice production. Emphasis should be placed on using smaller dosages of fertilizers because of the tremendous energy cost involved in their production. It is clear that the use of fertilizers could be significantly reduced if greater care were directed to the timing and the method of application. For example, placing fertilizer in the root zone rather than "top dressing" can double the effect of the fertilizer (IRRI, 1975). Small "mudballs" containing fertilizer can be placed by hand or by applicator about 10 centimetres into the soil immediately following transplanting or at a similar stage in the case of broadcast rice. Placing the fertilizer in the ground prevents microbial oxidation, thereby reducing the loss of nitrogen, potassium and phosphorus. The mudball technique is now being used in many parts of the Philippines, India, Laos and Thailand.

Time of fertilizing is just as important as method of application. Traditionally, fertilizers were applied only at the initial seeding stage. It is now clear that split applications can increase the yield using lower amounts of fertilizer. Simple dose applications of large amounts of fertilizer, e.g. 120 kg per hectare, may produce lower yields than multiple dose applications of lesser amounts of fertilizer. It is difficult to generalize from one region to another as far as method and timing of fertilizer application is concerned, but it can be safe to assume that in most cases moderate well-timed dosage applications produce higher yields than large-scale single-dose applications presently being employed with the "Green Revolution" method of rice production (IRRI, 1976).

With the introduction of high yielding hybrid varieties of rice, came the intensification of irrigation throughout the tropics. Water availability generally has been the main limiting factor to higher rice production in the tropics (Borgstrom, 1969 and Hogue, 1968). Since the early sixties, both above and below ground irrigation projects have doubled the amount of irrigable land (Brown, 1970). It has been speculated that the amount of irrigable land will double again by 1985 (Borgstrom, 1973). Lift-style irrigation is very energy intensive and although modern large-scale irrigation has become an important component of the agricultural strategy of the less developed world, very little consideration has been given to energetic and ecological ramifications. It is clear that minor irrigation schemes (i.e. gravity-type supplemented with tubewells) for the most part, have many advantages over large-scale dam and canal irrigation systems. For instance, it has been estimated that over half of the water diverted from rivers and streams in large-scale irrigation projects is lost to evaporation (Borgstrom, 1969). Waterlogging in the small-scale irrigation schemes is a far less serious problem, due to greater flexibility about quantity of water applied. More research should be directed towards more effective use of labour in the timing and application of irrigation water. A greater effort to build and maintain gravity-style irrigation canals should be exerted.

Further refinements of the auxiliary operations will not only lead to improvement in yield takeoffs and efficiency, but will significantly alleviate some of the environmental problems of the Third World such as soil erosion, salinization, eutrophication, chemical contamination and destruction of genetic diversity.

The methodology for evaluating modifications of rice production outlined in this chapter can be used to identify technological development options. Technologies, both intermediate and labour-intensive, can be moulded to the energy efficient transitional agricultural methods to minimize the use of cultural energy and maximize food yield.

Notes and Acknowledgements

This article is based on a paper originally presented at the 1980 annual meeting of the American Association for the Advancement of Science held in San Francisco and a paper entitled "Energy Constraints on Rice Production in the Developing World: An Environmental Perspective", International Journal of Environmental Studies (in press). The process models were designed by F. Alexander and the tables done by M. O'Dowd. Ms. O'Dowd and Ms. Shore-Freedman edited earlier versions of the manuscript.

References

Black, J. (1971): *Ann. Appl. Biol.,* 67, 272.

Borgstrom, G. (1973): *Focal Points: A Global Food Strategy.* Macmillan Press, New York.

Ibid. (1969): *Too Many: An Ecological Overview of Earth's Limitations.* Macmillan Press, New York.

Brown, L. (1970): *Seeds of Change: The Green Revolution and Development in the 1970s.* Praeger, New York.

Chambers, R. (Farmer, B. Ed.): *Green Revolution: Technology and Change in Rice-Growing Areas of Tamil, Nadu and Sri-Lanka.* Macmillan Press, New York.

Commoner, B. (1971): *The Closing Circle: Nature, Man and Technology.* Knopf, New York.

Freedman, S. (1978): *Modification of Traditional Rice Production Practices in the Developing World.* Doctorial Dissertation, University of California.

Ibid. (1979): *Energy Use Features.* (Fazzolare, R.; Smith, C. Ed.). Pergamon Press, Elmford, New York.

Ibid. Agroecosystems. (1980), 6, 129.

Hogue, A. (1968): *Costs and Returns: A Study of Irrigated Winter Crops.* East Pakistan Academy for Rural Development, Kotarbi.

International Rice Research Institute (1975): *The IRRI Annual Report.* Los Banos, Philippines, IRRI.

Ibid. (1976): *The IRRI Annual Report.* Los Banos, Philippines, IRRI.

Moowaw, J. and Curfs, H. (1972): Report on the Meeting of Experts on the Mechanization of Rice Production and Processing. UN Food and Agriculture Organization, Rome.

Matsubayashi, M. (1963): *Theory and Practice of Growing Rice.* Fuiji Publishing Co., Tokyo.

Pimentel, D.; Lynn, W.; MacReynolds, W.; Hewes, M.; and Rush, S. (1974): *Workshop on Research Methodologies for Studies of Energy, Food, Man and Environment, Phase One.* Centre For Environment Quality Management, Ithaca, New York.

Rappaport, R. (1971): *Sci. American,* 225, 117.

Triplett, G. Jr., Van Doren, D. (1977): *Sci. American,* 235, 28.